Advanced Materials Characterization

The book covers various methods of characterization of advanced materials commonly used in engineering including understanding of the working principle and applicability of devices. It explores the techniques implemented for advanced materials like superalloys, thin films, powders, nanocomposites, polymers, shape memory alloys, high entropy alloys, and so on. Major instruments covered include X-ray diffraction, near-field scanning optical microscopy Raman, X-ray photospectroscopy, ultraviolet-visible-near-infrared spectrophotometer, Fourier-transform infrared spectroscopy, differential scanning calorimeter, profilometer, and thermogravimetric analysis.

Features:

- Covers material characterization techniques and the development of advanced characterization technology
- Includes multiple length scale characterization approaches for a large variety of materials, from nano- to micron-scale, as well as their constraints
- Discusses advanced material characterization technology in the microstructural and property characterization fields
- Reviews both practical and theoretical explanations of approaches for characterizing microstructure and properties
- Offers fundamentals, basic instrumentation details, experimental approaches, analyses, and applications with case studies

This book is aimed at graduate students and researchers in materials science and engineering.

Advanced Materials Processing and Manufacturing
Series Editor: Kapil Gupta

The CRC Press Series in Advanced Materials Processing and Manufacturing covers the complete spectrum of materials and manufacturing technology, including fundamental principles, theoretical background, and advancements. Considering the accelerated importance of advances for producing quality products for a wide range of applications, the titles in this series reflect the state-of-the-art in understanding and engineering the materials processing and manufacturing operations. Technological advancements for enhancement of product quality, process productivity, and sustainability are on special focus including processing for all materials and novel processes. This series aims to foster knowledge enrichment on conventional and modern machining processes. Micro-manufacturing technologies such as micro-machining, micro-forming, and micro-joining, and Hybrid manufacturing, additive manufacturing, near net shape manufacturing, and ultra-precision finishing techniques are also covered.

Advanced Materials Characterization
Basic Principles, Novel Applications, and Future Directions
Ch Sateesh Kumar, M. Muralidhar Singh, and Ram Krishna

Thin-Films for Machining Difficult-to-Cut Materials
Challenges, Applications, and Future Prospects
Ch Sateesh Kumar and Filipe Daniel Fernandes

For more information about this series, please visit: www.routledge.com/Advanced-Materials-Processing-and-Manufacturing/book-series/CRCAMPM

Advanced Materials Characterization

Basic Principles, Novel Applications, and Future Directions

Ch Sateesh Kumar, M. Muralidhar Singh, and Ram Krishna

CRC Press
Taylor & Francis Group
Boca Raton London New York

CRC Press is an imprint of the
Taylor & Francis Group, an **informa** business

First edition published 2023
by CRC Press
6000 Broken Sound Parkway NW, Suite 300, Boca Raton, FL 33487-2742

and by CRC Press
4 Park Square, Milton Park, Abingdon, Oxon, OX14 4RN

CRC Press is an imprint of Taylor & Francis Group, LLC

ISBN: 9781032375106 (hbk)
ISBN: 9781032375113 (pbk)
ISBN: 9781003340546 (ebk)

DOI: 10.1201/9781003340546

Typeset in Times
by Newgen Publishing UK

Contents

Preface

The subject and the term "characterization" has become more common during the past few decades, although the main components of the process of characterization, such as mechanical testing, chemical analysis, etc., have been in use since the middle of the 19th century. The field of materials characterization brings together all these methods and techniques for the benefit of scientists, technicians, and experts concerned with quality assurance. The background of the variety of methods of characterization comes from several fundamental fields, like atomic physics, physical chemistry, mechanics of materials, and the science of materials itself as it has evolved over the last several decades. It is necessary that an engineer or technician trying to understand these methods is given the necessary background information by way of a brief revision so that the application is clearly understood. This has become particularly necessary as there are an extremely large number of different methods in use today, and many of them are quite sophisticated, requiring a good knowledge of the relevant fundamentals. An introduction to material characterization has been designed with the above need in mind. The book summarizes the various methods of characterization of materials commonly used in engineering (metals and alloys, ceramics, and polymers). Mostly, the books on materials discuss prominently the material properties and processing techniques. However, the material characterization techniques are only discussed briefly with minimal insights into the instrumentation and technical details of measuring devices and techniques. Thus, this book is focused on providing a comprehensive understanding of the working principle and applicability of devices implemented for the advanced characterization of materials. This book would explore the material characterization techniques implemented for advanced materials like superalloys, thin films, powders, nanocomposites, polymers, shape memory alloys, high entropy alloys, and so on.

Acknowledgments

To conclude, we would like to acknowledge the significant help that we have received from our colleagues, students, and family during the preparation of this book. We would like to acknowledge the sincere support from Prof. Kapil Gupta and Prof. Gagandeep Singh throughout the course of writing this book. We would also like to thank our departments – the Aeronautics Advanced Manufacturing Centre, Bilbao, Spain, Department of Mechanical Engineering, Madanapalle Institute of Technology & Science, India, and Department of Mechanical and Industrial Engineering Technology – for their motivation and support.

About the Authors

Ch Sateesh Kumar completed his PhD from the National Institute of Technology, Rourkela. He worked as Visiting Assistant Professor in Thapar Institute of Engineering and Technology. Later he joined the Advanced Materials Group at Czech Technical University, Prague as Postdoctoral Researcher. Presently he is working as Senior Researcher at the Aeronautics Advanced Manufacturing Centre, Bilbao, Spain in the Department of Mechanical Engineering, Madanapalle Institute of Technology & Science. He is also working as Senior Research Associate in the Department of Mechanical and Industrial Engineering Technology, University of Johannesburg. His research interests include machining, surface modification techniques, coating deposition and characterization, and tribology. He has published 21 research articles in highly reputed journals and also presented his work in 7 renowned conferences. He has also extended his services as a reviewer in many reputed journals.

M. Muralidhar Singh completed his PhD from The National Institute of Engineering, Visvesvaraya Technological University, Karnataka in 2017. He worked as Senior Assistant Professor in the Department of Mechanical Engineering at Madanapalle Institute of Technology & Science. He has served as Post Graduate Trainee at National Aerospace Lab (NAL), Bangalore. He is working as Assistant Professor in the Department of Mechanical Engineering at RV Institute of Technology and Management®, Bangalore. His research interests include nanomaterials, nanotechnology, hybrid thin films, lens for various optical instruments, wearable devices, sensor, and consumer goods. He has 13 years of research experience in several advanced materials fabrication and characterization techniques. He has published 24 research articles in the field of material science engineering, nanomaterials, composites and machining in highly reputed journals and also presented his research work in 8 renowned conferences. He has two granted Indian patents. He is actively reviewing manuscripts in top reputed journals.

Ram Krishna, after getting his Master's degree in Materials and Metallurgical Engineering from the Indian Institute of Technology Kanpur, continued his studies at the University of Leicester in the United Kingdom, where he got his PhD in 2011. He held postdoctoral positions at the University of Manchester in the United Kingdom and North Carolina State University in Raleigh, North Carolina. His research includes a wide range of materials for high temperatures applications and understanding their behavior and performance through mechanical testing at high temperatures, as well as advanced characterization techniques, as evidenced by his h-index of 12 and designation as co-inventor of 13 patents. He has authored or co-authored more than 30 research papers that have appeared in journals and conference proceedings. His current research focuses on three-dimensional (3D) printing and materials for 3D printing applications.

Introduction

This book is aimed at bridging the gap between the advancing techniques, applications, and fundamental concepts that have developed over the past few years. It has a broad scope of use and is intended to serve as a vehicle for developing, applying, and improving advancing technologies in a practical manner. As part of an advanced level course on the analysis of materials, this book covers the fundamental concepts, advanced techniques, and applications related to material characterization.

A comprehensive book covering state-of-the-art techniques in the field of material characterization, this book can be used as a single volume reference book. There are many chapters in this book that discuss microstructural characterization techniques that are commonly used in industry. In addition, there are a few chapters that discuss advanced techniques for characterizing microstructures and properties, which are also covered in this book with the applications.

The instruments will have an elaborative description of the scientific construction, working principle, and applicability of the discussed principle. The few instruments that will be discussed are as follows: X-ray diffraction, near-field scanning optical microscopy Raman, X-ray photospectroscopy, ultraviolet- visible-near-infrared spectrophotometer, Fourier-transform infrared spectroscopy, differential scanning calorimeter, profilometer, and thermogravimetric analysis. The discussion on the instruments will based on working principle, optics involved construction, sample standards, type of material that can be tested, and mounting methods. This book will be useful for undergraduate, postgraduate, and PhD scholars and researchers.

As often as possible, theories are presented to help the reader gain a better understanding of the fundamental principles of characterization techniques. An effort is made to demonstrate that the theories are legitimate. Throughout the book, technical aspects of specimen preparation and instrument use are discussed. This is in order to assist the reader in understanding both what to avoid and what should be followed in practice. The interpretation and analysis of the characterization outputs must be viewed as even more significant than the actual technical skills of characterization itself for most engineers and scientists.

DOI: 10.1201/9781003340546-1

It has become an essential part of today's scientific world to be able to interpret and analyze data, and this book provides the basis for this. As a means of assisting readers to make sense of and interpret the findings of an analysis, this book provides several examples for each of the techniques that are discussed in the book so that they can better understand them.

1 Introduction to Material Characterization

The term "characterization" has become more common during the last few decades, although the main components of the process of characterization, such as mechanical testing, chemical analysis, etc., have been in use since the middle of the 19th century. The field of materials characterization brings together all these methods and techniques for the benefit of scientists, technicians, and experts concerned with quality assurance. The background of the variety of methods of characterization comes from several fundamental fields, like atomic physics, physical chemistry, mechanics of materials, and the science of materials itself as it has evolved over the last several decades. An engineer or technician trying to understand these methods must be given the necessary background information by way of a brief revision so that the application is clearly understood. This has become particularly necessary as there are an extremely large number of different methods in use today, and many of them are quite sophisticated, requiring a good knowledge of the relevant fundamentals. An introduction to material characterization has been designed with the above need in mind.

1.1 APPLICATION OF ADVANCED CHARACTERIZATION TECHNOLOGIES

The use of advanced characterization technologies depends upon the type and the accuracy of information needed from the investigation. One such example can be the level of magnification obtained from different microscopy techniques. So, it is necessary to understand the characterization technique to finalize its application. The characterization techniques are used for studying surface morphology, cross-sectional morphology, elemental analysis, phase analysis, compositional changes, chemical bonding, and so on (Brzezinka et al., 2019; Das et al., 2016; Dosbaeva et al., 2015; Jiang et al., 2006; Kondo et al., 2020; Miletić et al., 2014; Thakur et al., 2015). The characterization techniques can be categorized basically into microscopy, spectroscopy, and macroscopic testing. However, it is collectively agreed that microscopic and spectroscopic techniques can be considered advanced characterization techniques (Wikipedia contributors, 2022a).

DOI: 10.1201/9781003340546-2

1.1.1 Microscopy

One of the most important characterization techniques is microscopy which finds its application in different fields of science and engineering (Hermans et al., 2018; McMullan, 1995). Microscopy is a technique that is used to study the surface and sub-surface structure of materials. However, microscopy can be subdivided into various other microscopic techniques depending on the source such as photons, electrons, ions, or physical cantilever probes that are used for gathering information about the surface of the sample specimen. Some examples of microscopy are optical microscopy, scanning electron microscopy, transmission electron microscopy, atomic force microscopy, and so on.

1.1.2 Spectroscopy

Spectroscopy is another important class of material characterization techniques that depends upon certain principles, such as X-ray diffraction, electron transmission, electron scattering, photoelectric effect, and so on, to give information about the chemical composition, chemical phases, chemical bonding, element mapping, crystal structure, grain orientation, etc. Some of the common examples of spectroscopy techniques are X-ray diffraction, X-ray photoelectron spectroscopy, energy dispersive spectroscopy (EDS), electron backscatter diffraction, electron energy loss spectroscopy, Raman spectroscopy, and so on (Hawkes & Spence, 2019; Kirk, 2017; Wikipedia contributors, 2022a, 2022b).

1.2 CHARACTERIZATION INSTRUMENTS

The characterization instruments are named after the principle they follow for getting information about the material whether it is a highly magnified image of a surface and a cross-section, or details about the crystal structure, chemical bonding, chemical composition, and chemical phases present in any material specimen. Different advanced characterization instruments would be discussed in detail in the subsequent chapters. In this section, we would understand the technology of some of the important microscopic and spectroscopic instruments that would be discussed later. An optical microscope is one of the common characterization instruments present in any laboratory. It is used for preliminary examination of getting magnified images of any material sample. An example is a measurement of cutting tool flank wear after machining on a stereo zoom optical microscope (Çalişkan et al., 2013; Vera et al., 2011). However, to get higher magnification, electron microscopy is very advantageous. These instruments use the principle of electron scattering, absorption, or transmission when a high-energy beam of electrons strikes the material sample under investigation. Examples of instruments following this principle are scanning electron microscope (SEM), field emission scanning electron microscope (FESEM), and transmission electron microscope (TEM) (Britannica, n.d.; Photometrics, n.d.; McMullan, 1995). Figure 1.1 shows a Zeiss-made (model number: EVO 18) SEM. The advantage of these electron micron microscopes is that they can provide extra information about the specimen such as chemical composition, phase transformation, crystal structure,

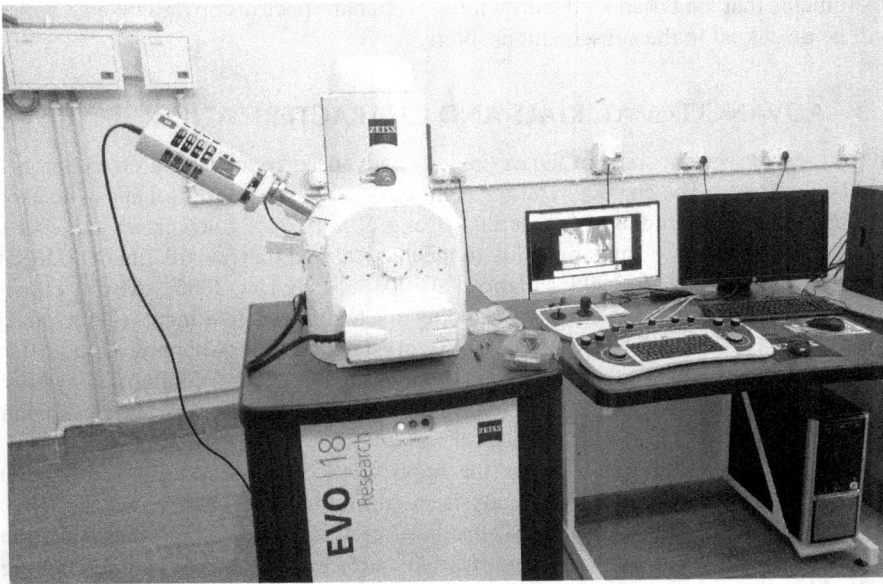

FIGURE 1.1 Zeiss-made EVO 18 scanning electron microscope. (https://en.wikipedia.org/wiki/File:CETB_Scanning_Electron_Microscope.jpg#filelinks. This work is licensed under the Creative Commons Attribution-ShareAlike 4.0 License.)

etc., in combination with attachments like EDS, and electron energy loss spectroscopy (EELS), etc. (Brzezinka et al., 2019; Wikipedia contributors, 2022b).

As discussed earlier that spectroscopic instruments provide useful information regarding crystal structure, phase transformation, chemical composition, etc., by following some principles which include the behavior of materials when they interact with different signals like X-rays. One such example is the X-ray diffractometer which uses the principle of X-ray diffraction. The multi-purpose X-ray diffractometers are all equipped with pre-aligned, quick-change X-ray modules that allow the user to switch beam paths effortlessly. These instruments help us obtain intensity data of a diffracted X-ray beam as a function of angle to satisfy Bragg's law under the condition of X-rays of known wavelength. Another example of one of the widely used spectroscopy techniques is X-ray photoelectron spectroscopy (XPS) which is a surface susceptible quantitative spectroscopic technique based on the photoelectric effect, used to identify the elements that may be present in a material or layer at the surface, as well as to determine their chemical state, the respective electronic structure and the density of electronic states in the different materials. XPS is the most commonly used measurement technique for analyzing various materials, showing not only which elements are present, but also the other elemental materials associated with the material. This characterization technique can be used to create line profiles of elemental composition on the surface or in conjunction with ion beam etching for depth profiling. These are some of the examples of microscopy and spectroscopy instruments. There are many other instruments used for advanced characterization instruments that are based on the principles of Raman spectroscopy, EDS, etc., which will be discussed in the subsequent chapters.

1.3   ADVANCED MATERIALS AND CHARACTERIZATION

In the present section, some of the examples of advanced characterization techniques will be discussed. The characterization techniques are powerful tools that can be used to study the microstructure of materials (Deng et al., 2017; Lauwers et al., 2008), phase transformation of materials due to mechanical and thermal treatments (Degen et al., 2014; Höling et al., 2005; Miletić et al., 2014; Shi & Liu, 2006), existing chemical bonding and changes in chemical bonding due to different treatments (Dobrzański & Żukowska, 2011; Dosbaeva et al., 2015), crystal structure (Chang & Duh, 2016; Mayrhofer et al., 2008), grain size (Lee, 1983; Wang et al., 2016), failure analysis of materials (Mulligan & Gall, 2005; Nohava et al., 2015; Rinaldi et al., 2018; Thakur et al., 2015), etc.

In this regard, Figure 1.2 shows the application of advanced characterization techniques for performing failure analysis in cutting tools (Kumar & Patel, 2018). Scanning electron microscopy (SEM) has been used to generate high-magnification images of the different wear zones on the rake surface of the mixed ceramic cutting tool which illustrates abrasion, adhesion, attrition, and chipping wear. Further, the EDS has been used to perform element analysis of the adhesion area and XRD has been performed to validate the oxidation of tool, workpiece, and coating material. From the above example, it is evident that the material characterization techniques are very helpful in finding and validating information related to material composition and phase transformation and at the same time generating magnified images with a high depth of field for detailed surface topographical analysis.

1.4   SUMMARY

Material characterization is an imperative part of materials science research that helps in studying and investigating the properties and structure of materials. These characterization techniques have sufficiently become advanced with the need for better and more accurate information regarding the properties and structure of different materials. The advanced material characterization techniques can be classified into microscopy and spectroscopy techniques. Microscopy is used to study the surface and subsurface structure of materials. However, microscopy can be subdivided into various other microscopic techniques depending on the source such as photons, electrons, ions, or physical cantilever probes that is used for gathering information about the surface of the sample specimen. On the contrary, spectroscopy defines the material characterization techniques that depend on certain principles, such as X-ray diffraction, electron transmission, electron scattering, photoelectric effect, and so on, to give information about the chemical composition, chemical phases, chemical bonding, element mapping, crystal structure, grain orientation, etc. These techniques in combination can act as a powerful tool for investigating the chemical composition, crystal structure, phase transformation, etc., in highly magnified regions.

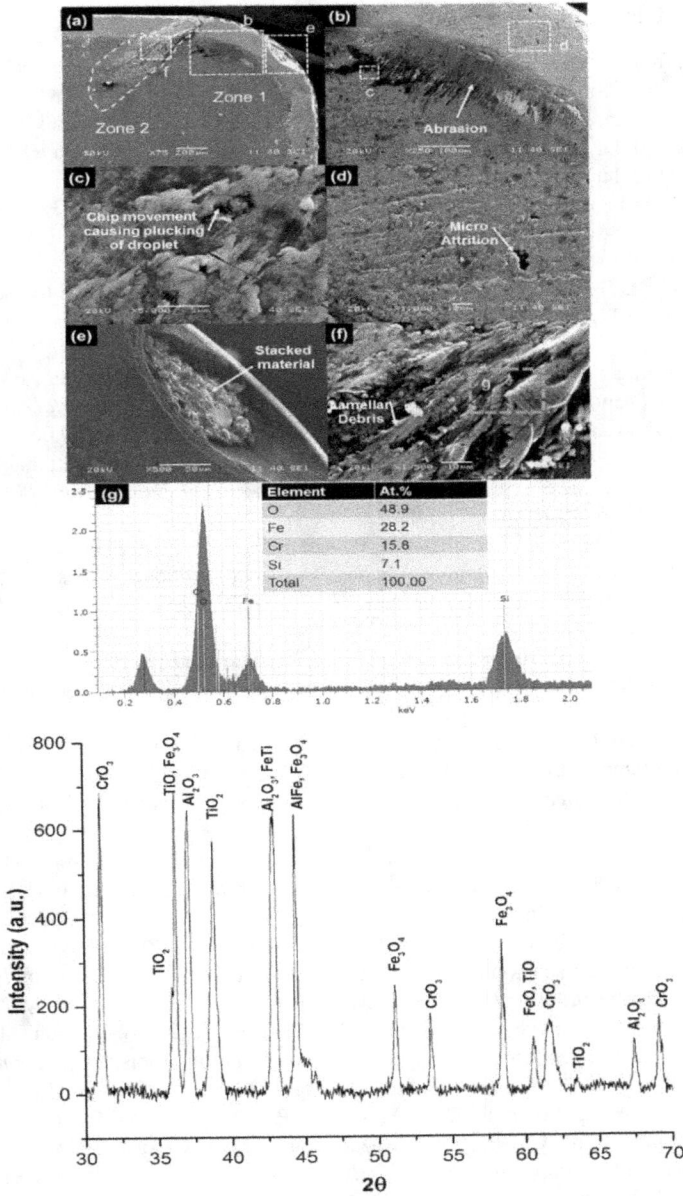

FIGURE 1.2 SEM micrographs and XRD phase analysis of the rake surface of monolayered AlCrN coated cutting tool after machining. (For details, see "Kumar, C. S., & Patel, S. K. (2018). Performance analysis and comparative assessment of nano-composite TiAlSiN/TiSiN/TiAlN coating in hard turning of AISI 52100 steel. *Surface and Coatings Technology*, *335*(September 2017), 265–279. https://doi.org/10.1016/j.surfcoat.2017.12.048. Reprinted with permission from Elsevier.)

REFERENCES

Britannica. (n.d.). Retrieved August 16, 2022, from www.britannica.com/technology/transmiss ion-electron-microscope

Brzezinka, T., Rao, J., Paiva, J., Kohlscheen, J., Fox-Rabinovich, G., Veldhuis, S., & Endrino, J. (2019). DLC and DLC-WS2 coatings for machining of aluminium alloys. *Coatings, 9*(3), 192. https://doi.org/10.3390/coatings9030192

Çalişkan, H., Erdoğan, A., Panjan, P., Gök, M. S., & Karaoğlanli, A. C. (2013). Micro-abrasion wear testing of multilayer nanocomposite tialsin/tisin/tialn hard coatings deposited on the aisi h11 steel. *Materiali in Tehnologije, 47*(5), 563–568.

Chang, C. C., & Duh, J. G. (2016). Duplex coating technique to improve the adhesion and tribological properties of CrAlSiN nanocomposite coating. *Surface and Coatings Technology*. https://doi.org/10.1016/j.surfcoat.2016.11.032

Das, P., Anwar, S., Bajpai, S., & Anwar, S. (2016). Structural and mechanical evolution of TiAlSiN nanocomposite coating under influence of Si3N4 power. *Surface and Coatings Technology, 307*, 676–682. https://doi.org/10.1016/j.surfcoat.2016.09.065

Degen, F., Klocke, F., Bergs, T., & Ganser, P. (2014). Comparison of rotational turning and hard turning regarding surface generation. *Production Engineering, 8*(3), 309–317. https://doi.org/10.1007/s11740-014-0530-6

Deng, Y., Tan, C., Wang, Y., Chen, L., Cai, P., Kuang, T., Lei, S., & Zhou, K. (2017). Effects of tailored nitriding layers on comprehensive properties of duplex plasma-treated AlTiN coatings. *Ceramics International, 43*(12), 8721–8729. https://doi.org/10.1016/j.ceram int.2017.03.209

Dobrzański, L. A., & Żukowska, L. W. (2011). Structure and properties of gradient PVD coatings deposited on the sintered tool materials materials. *Tool Materials Journal of Achievements in Materials and Manufacturing Engineering, 442*(2), 115–139.

Dosbaeva, G. K., El Hakim, M. A., Shalaby, M. A., Krzanowski, J. E., & Veldhuis, S. C. (2015). Cutting temperature effect on PCBN and CVD coated carbide tools in hard turning of D2 tool steel. *International Journal of Refractory Metals and Hard Materials, 50*, 1–8. https://doi.org/10.1016/j.ijrmhm.2014.11.001

Hawkes, P. W., & Spence, J. C. H. (2019). *Springer Handbook of Microscopy*, ISBN: 978-3-030-00068-4. Springer. https://doi.org/10.1007/978-3-030-00069-1

Hermans, J., Osmond, G., Van Loon, A., Iedema, P., Chapman, R., Drennan, J., Jack, K., Rasch, R., Morgan, G., Zhang, Z., Monteiro, M., & Keune, K. (2018). Electron microscopy imaging of zinc soaps nucleation in oil paint. *Microscopy and Microanalysis, 24*(3), 318–322. https://doi.org/10.1017/S1431927618000387

Höling, A., Hultman, L., Odén, M., Sjölén, J., & Karlsson, L. (2005). Mechanical properties and machining performance of Ti1-xAlxN-coated cutting tools. *Surface and Coatings Technology, 191*(2–3), 384–392. https://doi.org/10.1016/j.surfcoat.2004.04.056

Jiang, W., More, A. S., Brown, W. D., & Malshe, A. P. (2006). A cBN-TiN composite coating for carbide inserts: Coating characterization and its applications for finish hard turning. *Surface and Coatings Technology, 201*(6), 2443–2449. https://doi.org/10.1016/j.surfc oat.2006.04.026

Kirk, T. L. (2017). A review of scanning electron microscopy in near field emission mode. In *Advances in Imaging and Electron Physics* (1st ed., Vol. 204). Elsevier Inc. https://doi.org/10.1016/bs.aiep.2017.09.002

Kondo, T., Inoue, H., & Minoshima, K. (2020). Thickness dependency of creep crack propagation mechanisms in submicrometer-thick gold films investigated using in situ FESEM and EBSD analysis. *Materials Science and Engineering A, 790*, 139621. https://doi.org/10.1016/j.msea.2020.139621

Kumar, C. S., & Patel, S. K. (2018). Performance analysis and comparative assessment of nano-composite TiAlSiN/TiSiN/TiAlN coating in hard turning of AISI 52100 steel. *Surface and Coatings Technology, 335*(September 2017), 265–279. https://doi.org/10.1016/j.surfcoat.2017.12.048

Lauwers, B., Brans, K., Liu, W., Vleugels, J., Salehi, S., & Vanmeensel, K. (2008). Influence of the type and grain size of the electro-conductive phase on the Wire-EDM performance of ZrO2 ceramic composites. *CIRP Annals–Manufacturing Technology, 57*(1), 191–194. https://doi.org/10.1016/j.cirp.2008.03.089

Lee, M. (1983). High temperature hardness of tungsten carbide. *Metallurgical Transactions A, 14*(8), 1625–1629. https://doi.org/10.1007/BF02654390

Mayrhofer, P. H., Music, D., Reeswinkel, T., Fuß, H. G., & Schneider, J. M. (2008). Structure, elastic properties and phase stability of Cr1-xAlxN. *Acta Materialia, 56*(11), 2469–2475. https://doi.org/10.1016/j.actamat.2008.01.054

Miletić, A., Panjan, P., Škorić, B., Čekada, M., Dražič, G., & Kovač, J. (2014). Microstructure and mechanical properties of nanostructured Ti-Al-Si-N coatings deposited by magnetron sputtering. *Surface and Coatings Technology, 241*, 105–111. https://doi.org/10.1016/j.surfcoat.2013.10.050

McMullan, D. (1995). Scanning electron microscopy 1928–1965. *Scanning*, 17, 175–185. https://doi.org/10.1002/sca.4950170309

Mulligan, C. P., & Gall, D. (2005). CrN-Ag self-lubricating hard coatings. *Surface and Coatings Technology, 200*(5–6), 1495–1500. https://doi.org/10.1016/j.surfcoat.2005.08.063

Nohava, J., Dessarzin, P., Karvankova, P., & Morstein, M. (2015). Characterization of tribological behavior and wear mechanisms of novel oxynitride PVD coatings designed for applications at high temperatures. *Tribology International, 81*, 231–239. https://doi.org/10.1016/j.triboint.2014.08.016

Photometrics, Inc. (n.d.). Retrieved August 13, 2022, from https://photometrics.net/field-emiss ion-scanning-electron-microscopy-fesem/

Rinaldi, S., Caruso, S., Umbrello, D., Filice, L., Franchi, R., & Del Prete, A. (2018). Machinability of Waspaloy under different cutting and lubri-cooling conditions. *International Journal of Advanced Manufacturing Technology, 94*(9–12), 3703–3712. https://doi.org/10.1007/s00170-017-1133-0

Shi, J., & Liu, C. R. (2006). On predicting chip morphology and phase transformation in hard machining. *International Journal of Advanced Manufacturing Technology, 27*(7–8), 645–654. https://doi.org/10.1007/s00170-004-2242-0

Thakur, A., Gangopadhyay, S., & Mohanty, A. (2015). Investigation on some machinability aspects of inconel 825 during dry turning. *Materials and Manufacturing Processes, 30*(8), 1026–1034. https://doi.org/10.1080/10426914.2014.984216

Vera, E. E., Vite, M., Lewis, R., Gallardo, E. A., & Laguna-Camacho, J. R. (2011). A study of the wear performance of TiN, CrN and WC/C coatings on different steel substrates. *Wear, 271*(9–10), 2116–2124. https://doi.org/10.1016/j.wear.2010.12.061

Wang, Q., Zhou, F., & Yan, J. (2016). Evaluating mechanical properties and crack resistance of CrN, CrTiN, CrAlN and CrTiAlN coatings by nanoindentation and scratch tests. *Surface and Coatings Technology, 285*, 203–213. https://doi.org/10.1016/j.surfcoat.2015.11.040

Wikipedia contributors. (2022a). *Characterization (materials science) – Wikipedia, The Free Encyclopedia.* https://en.wikipedia.org/w/index.php?title=Characterization_(materi als_science)&oldid=1089618338

Wikipedia contributors. (2022b). *Transmission electron microscopy – Wikipedia, The Free Encyclopedia.* https://en.wikipedia.org/w/index.php?title=Transmission_electron_mic roscopy&oldid=1103921398

2 X-Ray Diffraction (XRD)

X-ray diffraction (XRD) is based on the constructive interference of monochromatic X-rays and a crystalline sample. These X-rays are generated by a cathode ray tube, filtered to produce monochromatic radiation, collimated for the concentrate, and directed toward the sample. The interaction of the incident beams with the sample results in constructive interference (and a diffracted beam) if the conditions satisfy Bragg's law ($n\lambda = 2d \sin \theta$). This law relates the wavelength of electromagnetic radiation to the diffraction angle and lattice spacing in a crystalline sample. These diffracted X-rays are then detected, processed, and counted. By scanning the sample over a range of 2θ-angles, all possible diffraction directions of the grating should be obtained due to the random orientation of the powdered material. Converting the diffraction peaks to d-spacings allows identification of the mineral since each mineral has a unique set of d-spacings. This is usually achieved by comparing the d-spacings to standard reference patterns.

All diffraction methods rely on the generation of X-rays in an X-ray tube. These X-rays are directed at the sample, and the diffracted beams are collected. A key component of all diffraction methods is the angle between the incident and diffracted beams. Powder and single-crystal diffractions also differ in instrumentation.

XRD is a technique used in materials science to determine the atomic and molecular structure of materials. It involves irradiating a material sample with incident X-rays and then measuring the intensities and scattering angles of the X-rays scattered by the material. The intensity of the scattered X-rays is recorded as a function of the scattering angle, and the structure of the material is determined from the analysis of the position, angle, and intensities of the scattered intensity peaks. In addition, the standard positions of the atoms in the crystal can be measured, and it can be determined how the actual structure deviates from the ideal structure, for example, due to internal stresses or defects. The diffraction of X-rays, which is the focus of the XRD method, is a subset of general X-ray scattering phenomena. XRD, generally referred to as wide-angle X-ray diffraction (WAXD), falls under several methods that utilize elastically scattered X-ray waves, Other X-ray techniques based on elastic scattering include small-angle X-ray scattering (SAXS), in which X-rays are incident on the sample over a small angular range (0.1–100 typically). SAXS measures structural correlations on the order of a few nanometers or more (e.g., crystal

DOI: 10.1201/9781003340546-3

superstructures) and X-ray reflectivity, which measures the thickness, roughness, and density of thin films. WAXD covers an angular range of over 100.

2.1   CONSTRUCTION OF X-RAY DIFFRACTOMETER

X-rays can be produced by exciting target atoms by bombarding them with charged particles or by absorbing electromagnetic radiation. It contains a filament (wire) and an anode (target) housed in a vacuum enclosure. An electric current heats the filament and electrons are emitted. A high voltage of about 30–45 kV is applied between the filament and the anode, accelerating the electrons toward the anode. When the electrons hit the anode, they are decelerated, resulting in the emission of X-rays (see Figure 2.1).

XRD experiments use the characteristic K_α lines of the metal anticathode. To obtain a monochromatic radiation, the beam can be filtered through a material that selectively absorbs the CuK_β line. A Ni filter strongly absorbs the CuK_β peak. When the incident beam hits a sample, diffraction occurs at every possible 2θ orientation. The diffracted beam can be detected using a moving detector. In normal use, the counter is set to sample a range of 2θ values at a constant angular velocity. Typically, a 2θ range of 5–70°C is sufficient to cover the most useful part of the powder pattern (see Figure 2.2).

The X-ray diffractometer is a precision instrument with two freely rotating axes (ω and 2θ). With this instrument, we can obtain the intensity data of a diffracted X-ray beam as a function of angle, so that Bragg's law is satisfied under the condition of X-rays of known wavelength. The basic structure of the diffractometer is shown in Figure 2.3. Three components, the X-ray source (F), the sample holder (S), and the detector (G), are located on the circumference of a circle called the focusing circle. When the position of the X-ray source is fixed and the detector is placed on the 2θ-axis, a powder sample in the form of a flat plate is usually placed

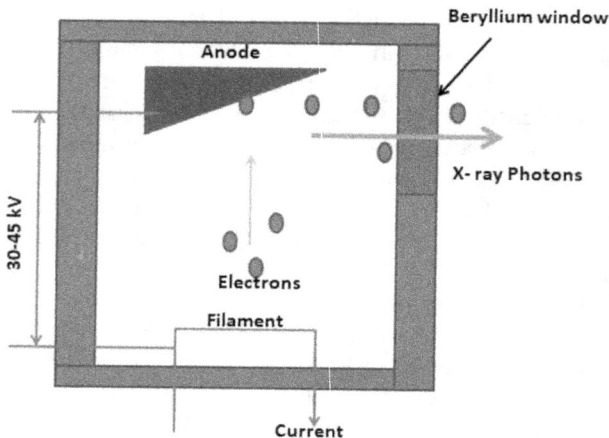

FIGURE 2.1   Schematic representation of X-rays.

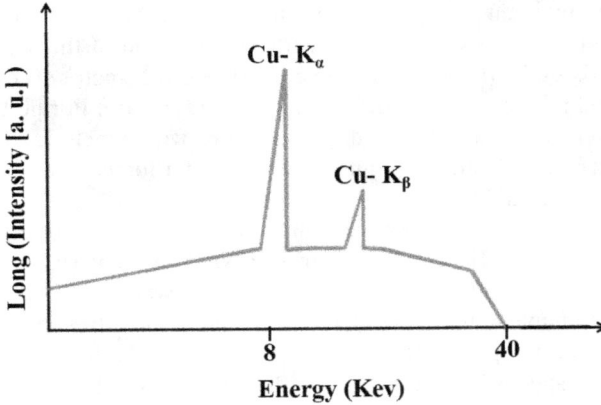

FIGURE 2.2   XRD source wavelength.

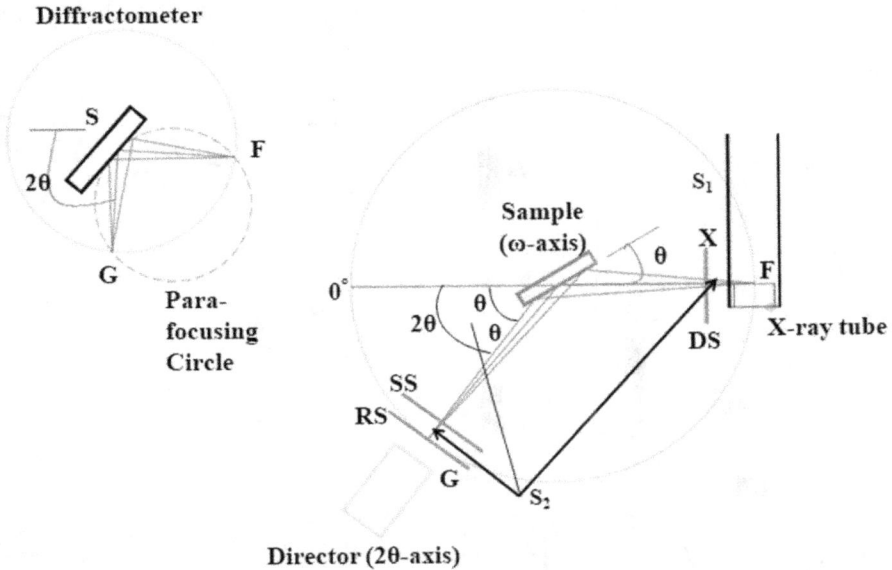

FIGURE 2.3   Construction of an X-ray diffractometer.

on the ω-axis, which corresponds to the center of the diffractometer. The line focal point on the X-ray tube target is aligned parallel to the ω-axis of the diffractometer. The main reason for using a flat plate sample is to take advantage of the focusing geometry to effectively detect the intensity of weak diffracted beams. During measurements, the 2θ-axis rotates twice as much as the ω-axis, which is why we often refer to it as the θ, 2θ-scan to maintain the experimental condition that the angle between the sample plane and the direction of the incident X-ray beam is equal to that of the direction of the diffracted beam, relative to the propagation direction of the incident X-ray beam. In other words, the direction of the normal to the specimen plane should coincide with the direction of the scattering vector q=s-s0, which is defined by the difference between the vector s0 of the incident X-ray beam and the vectors s of the diffracted X-ray beam. Furthermore, a circle through the points F (focal point on the target), S (diffractometer), and G (the focal point of the diffracted beam) in Figure 2.3 is called a focusing circle or Rowland circle (Waseda et al., n.d.).

To minimize angular dispersion and improve spatial resolution for the incident X-ray beam as well as the diffracted X-ray beam, some slit systems are inserted into the X-ray beam path. We also use a Soller slit, which consists of a series of closely spaced thin metal plates parallel to the plane of the diffractometer circle, and is designed to limit the perpendicular dispersion of both the incident and diffracted X-ray beams. As shown in Figure 2.3, the divergence slits (DS) and the scattering slits (SS) are adjusted to limit the horizontal dispersion of the incident and diffracted X-rays, and the receiving slit (RS) in front of the detector is used to determine the spatial resolution. The important feature of a diffractometer is not only the limitation of dispersion by DS and SS but also the focusing of the diffracted X-ray beam from

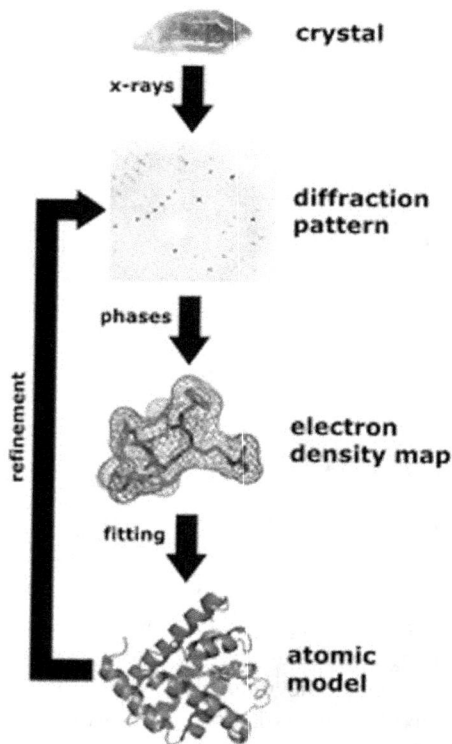

FIGURE 2.4 Structure for a molecule by X-ray diffractometer. (https://en.wikipedia.org/wiki/X-ray_crystallography#Early_scientific_history_of_crystals_and_X-rays).

powder samples by RS. This collimation and focusing principle is called parafocusing (Junhui et al., 2011). As shown in Figure 2.3, the position of RS in front of the detector always coincides with a parafocusing point in the diffractometer, making the intensity measurement effective and the spatial resolution better.

The oldest and most accurate method of XRD, in which an X-ray beam strikes a single crystal, producing scattered rays that create a diffraction pattern of spots when they strike a film or other detector; the strength and angles of these rays are recorded as the crystal is gradually rotated. Each spot is called a reflection because it corresponds to the reflection of X-rays from a series of uniformly distributed planes in the crystal. For single crystals of sufficient purity and regularity, XRD data can be used to determine the average chemical bond lengths and angles with an accuracy of a few thousandths of an angstrom and a few tenths of a degree, respectively (see Figure 2.4). The atoms in a crystal are not static, but oscillate about their mean position, usually by less than a few tenths of an angstrom. X-ray crystallography allows the magnitude of these oscillations to be measured (Florian et al., 2007).

2.2 WORKING PRINCIPLE

XRD is based on the constructive interference of monochromatic X-rays and a crystalline sample. These X-rays are produced by a cathode ray tube, filtered to produce monochromatic radiation, collimated to focus it, and directed at the sample. When a monochromatic X-ray beam strikes a crystal, the atomic electrons in the crystal are set into vibration. They have the same frequency as that of the incident radiation and are accelerated. If the wavelength of the incident radiation is large compared to the dimensions of the crystal, then the emitted X-rays will be in phase with each. However, since the atomic dimensions are almost equal to the wavelength of the X-rays, the rays emitted by the electrons are out of phase. This radiation can constructively or destructively interfere and produce a diffraction pattern (i.e., maxima and minima) in certain directions (Newman, n.d.) Bragg's law of diffraction To explain the diffraction of X-rays, W.L. Bragg considered XRD from a crystal as a problem of reflection of X-rays from the atomic planes of the crystal according to the reflection laws, as shown in Figure 2.5. Consider a series of parallel atomic planes of the crystal with Miller indices such that the distance between the two consecutive planes is d. Let a parallel beam of monochromatic X-rays of wavelength λ be incident on the plane at a grazing angle θ such that the incident rays lie in the plane of the paper (Waseda et al., n.d.).

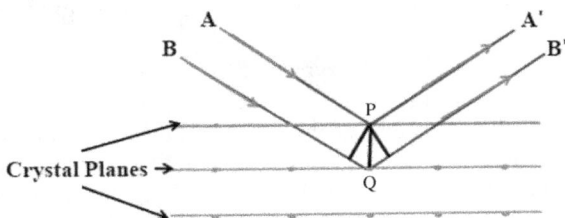

FIGURE 2.5 X-ray diffraction principle.

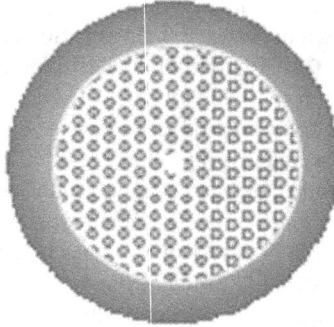

FIGURE 2.6   Laue pattern of ZnS crystal.

AP and BQ be two parallel incident rays reflected from points P and Q on the crystal planes and traveling at PA' and QB', respectively, as shown in Figure 2.5. When the path difference between APA' and BQB' is an integer multiple of λ, constructive interference occurs and a maximum is observed. Max von Laue, a German theoretical physicist, proposed in 1912 that. If a crystal consists of a regular and ordered arrangement of atoms, the planes of the crystal should have a spacing of the order of 1 Armstrong. And a crystal should behave like a neutral three-dimensional diffraction grating for the diffraction of X-rays. These also have a wavelength of about 1Å (Palmieri et al., 2011). It was therefore predicted that when an inhomogeneous X-ray beam falls on a crystal, a series of geometrically arranged spots, about the center of the incident beam, should be obtained in the diffraction pattern, which is on a photographic plate list on the other side of the crystal.

This was actually verified experimentally, and a definite arrangement of atoms in the crystal was found. The spots that appeared on the photographic plate when the experiment was performed with the ZnS crystal were called Laue spots. And the pattern of the spots on the photographic plate was called the Laue pattern. The Laue pattern of the ZnS crystal is shown in Figure 2.6.

2.3   DIFFRACTION THEORY

The instruments used for powder diffraction measurements have changed little from those developed in the late 1940s. The major difference in modern instruments is the use of minicomputers for control, data acquisition, and data processing. Figure 2.7 shows the geometry of the system.

The geometrical arrangement of a typical diffractometer system shows the source F, the Soller slits P and RP, the sample S, the divergence slit D, and the receiving slit R. The axis of the goniometer is at A. This geometrical arrangement is known as the Bragg–Brentano parafocusing system and is described by a diverging beam from a line source F falling on the sample S, being diffracted, and passing through a receiving slit R to the detector. The distances FA and AR are equal. The amount of divergence is determined by the effective focal length of the source and the aperture

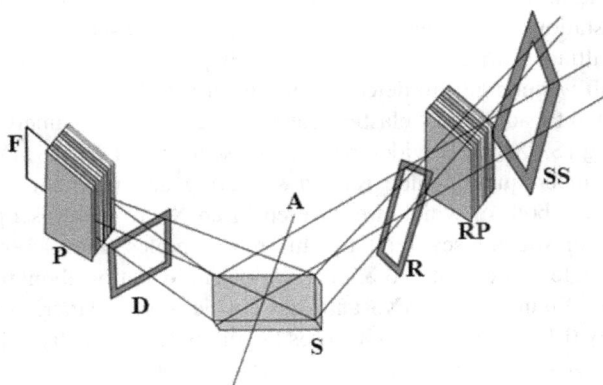

FIGURE 2.7 Geometry of the Bragg–Brentano diffractometer.

of the divergence slit D. The axial divergence is controlled by two sets of parallel plate collimators (Soller slits) P and RP, located between the focus and the sample and between the sample and the divergence slit, respectively (Mendenhall et al., 2015). Using the narrower divergence slit results in less sample coverage for a given diffraction angle, so lower diffraction angles can be achieved if the sample has a larger apparent surface area (resulting in larger values for d). Thus, the choice of divergence slit and matching scattering slit depends on the angular range to be covered. The decision of whether or not to enlarge the slit at a given angle is determined by the available intensity. A photon detector, usually a scintillation detector, is placed behind the scattering slit and converts the diffracted X-ray photons into voltage pulses. By synchronizing the sampling rate of the goniometer with the recorder, a diffractogram is obtained in which the 2θ degree is plotted against the intensity. A timer/scaler is also provided for quantitative work and is used to measure the integrated peak intensity of one or more selected lines from each analysis phase in the sample. A diffracted beam monochromator may also be used to improve the signal-to-noise ratio. The output of the diffractometer is a "powder chart," essentially a plot of intensity as a function of diffraction angle, which can be in the form of a strip chart or a printout from a computer graphics terminal (Brindley & Brown, 1981). The powder method owes its name to the fact that the sample is typically in the form of a microcrystalline powder, although, as mentioned earlier, any material consisting of an ordered arrangement of atoms will give a diffraction pattern. The possibility of using a diffraction pattern as a means of phase identification was recognized as early as 1935, but it was not until the late 1930s that a systematic procedure for deciphering the superimposed diffraction patterns was proposed by Hanawalt et al. (1938).

2.4 SCATTERING TECHNIQUES

In elastic scattering, the scattered X-rays have the same energy/wavelength as the incident X-rays, while in inelastic scattering, the scattered X-rays have a different energy/wavelength. One of the best-known elastic scattering techniques is X-ray diffraction/crystallography. These two techniques are subsets of X-ray scattering that use crystalline samples and require very high-energy hard X-rays (with an extremely small wavelength) to detect atomic-level detail. The X-ray scattering can be categorized into additional elastic scattering techniques, namely, small angle X-ray scattering (SAXS) and wide-angle X-ray scattering (WAXS) (Waseda et al., n.d.). Inelastic techniques include Raman scattering and resonant inelastic X-ray scattering (RIXS), both of which are covered in an X-ray spectroscopy document. X-rays can be elastically scattered by different samples at different angles. By looking at the angle of the scattered X-rays, various information about the sample can be obtained (see Figure 2.8). SAXS analyzes elastic X-ray scattering at very small angles, typically 0.1–10°, while WAXS uses larger angles, typically > 10°.

By detecting X-rays scattered at different angles, different sample information is obtained at different resolutions. SAXS has a nanoscale resolution (good for samples of 100–1 nm in size), while WAXS has an atomic resolution (1–0.1 nm). For this reason, SAXS can typically be used to analyze larger microstructures within a sample, while WAXS is more similar to XRD and can observe atomic details. SAXS and WAXS can be performed simultaneously and are easily interchangeable, requiring only that the detector and sample be moved closer or farther apart (see Figure 2.9).

Because SAXS/WAXS can be used on noncrystalline samples, it can be used to collect information on highly complex samples, including proteins and compound systems. They have been used to study drug delivery systems, formulations, phase behavior, and advanced materials development.

Proper sample preparation is one of the most important requirements for the analysis of powder samples by XRD. This statement is especially true for soils and clays that contain finely divided colloids that reflect X-rays poorly, as well as for other types of materials such as iron oxide coatings and organic materials that make characterization by XRD difficult. Sample preparation involves not only the proper treatment of samples to remove unwanted substances, but also appropriate techniques to obtain the desired particle size, orientation, thickness, and so forth. Several excellent books are available that address appropriate sample preparation techniques for clays and soils (Bish & Post 1994; Jackson, 1981; Moore & Reynolds, 1989).

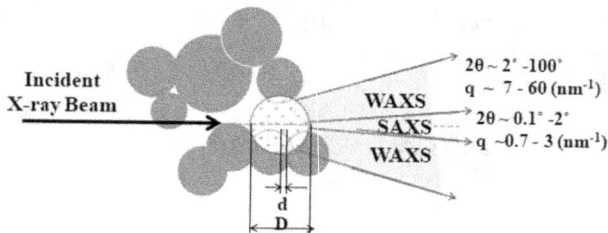

FIGURE 2.8 SAXS and WAXS achieve different resolutions, with WAXS measuring wider angles and achieving atomic resolution, and SAXS measuring very small angles and achieving nanoscale resolution.

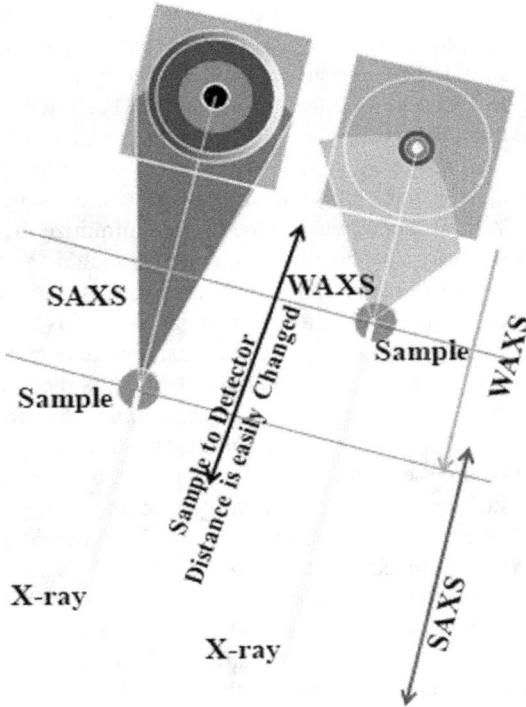

FIGURE 2.9 SAXS and WAXS can be done simultaneously by moving the detector closer or further away from the sample.

2.5 SIO$_2$ GLASS

Analysis of powders by XRD requires that they must be extremely fine-grained to achieve a good signal-to-noise ratio (and avoid variations in intensity), avoid spotting, and minimize preferred orientation. The comminution of powders into fine particles also ensures adequate particle participation in the diffraction process. The recommended size range is 1–5 μm (Klug & Alexander, 1974). For routine qualitative evaluation of mineral constituents, samples are usually ground and sieved through a 325 mesh (45 μm) sieve. Grinding is done either by hand grinding or in a mechanical mill. The effects of excessive grinding include lattice distortion and the possible formation of an amorphous layer (Beilby layer) outside the grains (Vogt & Williams, 2011). Depending on the type of crystallite orientation desired, two types of embeddings are generally used: Random embeddings are preferred when phases are to be identified in a sample. In this type of embedding, particles crushed to 1–5 μm are packed onto a sample holder on a flat surface to assume different orientations and ensure reflections from different planes. Oriented holders are used in the analysis of clay minerals, which rarely show strong diffraction effects from Bragg planes other than the (001). Generally, they are prepared by slurring the sample with distilled water. The water is then evaporated until the slurry is spread on a specimen holder (e.g., a glass slide or ceramic tile). The important factors in specimen preparation are described below. Sample characteristics also affect the quality of a powder pattern by either reducing intensities or distorting intensities. Preferred orientation or texture: texture means that the powder particles do not have a random shape, but a highly regular anisotropic shape, typically platelets or needles. During preparation, these are then preferentially aligned along the sample surface, causing massive changes in peak intensities. Various techniques can be used to minimize this effect. The most efficient method is to slurry in a highly viscous liquid such as nail polish, in which the random alignment is preserved as it dries. Alternatively, the anisotropic particle shape can be reduced by grinding in a ball mill, although this should be done with great care, as excessive grinding can easily reduce the particle size to the nanometer range and lead to amorphization (Higashi et al., 2010). It is recommended to try the effect of successive 5-minute grinding intervals to optimize the process on each specimen. For coatings or thin films, preferential alignment is often the desired effect. Crystallite size and strain: the width of a diffraction peak is reciprocal to the average crystallite size, the smaller the average crystallite size, the wider the reflections and the lower the absolute intensities. This effect is visible below an average crystallite size of less than 200 nm. Related to the broadening of the crystallite size is the strain broadening. It arises from the presence of defects in the crystals. Such elongation can be produced by substitution of the atoms of which it is composed, but also by special heat treatment. It is possible to distinguish strain broadening from size broadening since the angular dependence is much greater than in the latter case: rotating specimen holders improve the measurement statistics and thus give the best results. However, they are not available for all machines (Andreeva et al., 2011). The most serious error in specimen preparation is to fill the specimen holder too high or too low. Both lead to a significant shift of the peak positions, which can complicate the evaluation.

The three XRD patterns above were produced by three chemically identical forms of SiO_2. Crystalline materials such as quartz and cristobalite produce XRD patterns – quartz and cristobalite have two different crystal structures. The Si and O atoms are arranged differently, but both exhibit extensive atomic ordering. The difference in their crystal structure is reflected in their different diffraction patterns. The amorphous glass does not exhibit long-range atomic ordering and therefore produces only broad scattering features, as shown in Figure 2.10.

Each diffraction peak is attributed to scattering from a particular group of parallel atomic planes.

Miller indices (hkl) are used to identify the different atomic planes. The observed diffraction peaks can be assigned to atomic planes to aid in the analysis of the atomic structure and microstructure of a sample as shown in Figure 2.11 (Junhui et al., 2011). The (hkl)-plane of atoms intersects the unit cell at a/h, b/k, and c/l, the (220)-plane drawn to the right intersects the unit cell at ½a, ½b, and does not intersect the c-axis. When a plane is parallel to an axis, it is assumed to intersect the ∞-axis; therefore, its reciprocal is 0. The vector dhkl is drawn from the origin of the unit cell so that it intersects the crystallographic plane (hkl) at an angle of 90°. The direction of dhkl is the crystallographic direction. The crystallographic direction is expressed with [] in brackets, e.g. [220].

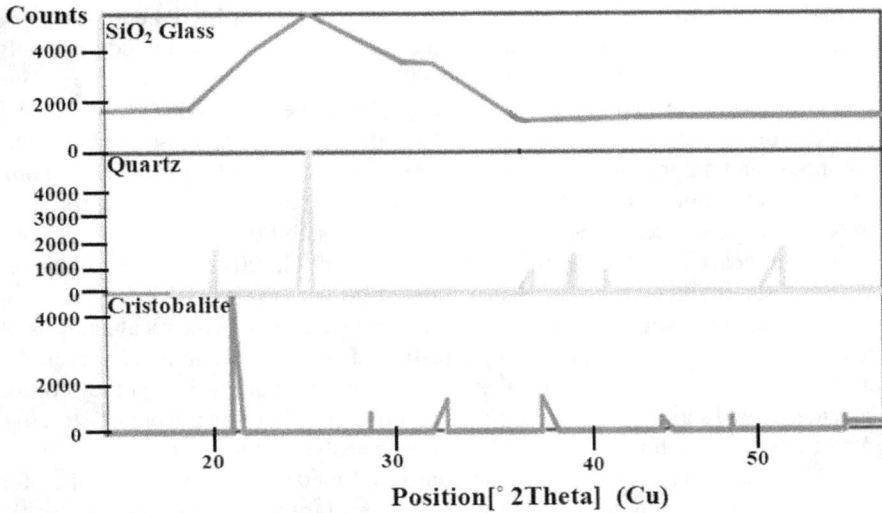

FIGURE 2.10 X-rays scatter from atoms in a material and therefore contain information about the atomic arrangement.

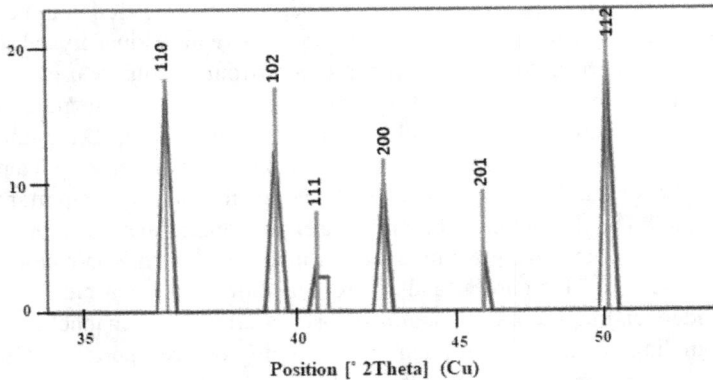

FIGURE 2.11 Diffraction pattern to treat a crystal as a collection of planes of atoms.

2.6 APPLICATIONS

The pharmaceutical industry is one of the most successful industries in the technology sector, and its ability to innovate has resulted in nearly 1,400 new chemical entities being launched as human therapeutics over the past 30 years. Despite this success, the research and development (R&D) process to successfully bring a drug to market remains challenging. Drug development is a risky and expensive process to increase demand for more available and affordable drugs (generics) and to propose adapted new drugs (Ivanisevic et al., 2010). Solid form screening, i.e., the preparation and analysis of various solid forms of Active Pharmaceutical Ingredient (API), has become an essential part of drug development. Selection criteria include not only pharmaceutically relevant properties such as therapeutic efficacy and processing characteristics, but also intellectual property (IP) issues (Aaltonen et al., 2009). A variety of screens can be performed, including polymorphic, salt, crystalline, amorphous, and amorphous dispersions. X-ray powder diffraction (XRPD) is commonly used at various stages of screening to identify and characterize new forms. It also helps evaluate other properties, such as physical stability and manufacturability, to select the best shape for development (Borghetti et al., 2012). These techniques were also used to obtain the crystal structures of three of the five polymorphs of m-aminobenzoic acid (Vogt & Williams, 2011). Forensic Science Chemical analysis of forensic "samples" usually means identification and/or comparisons. However, the samples differ from most samples encountered in other situations in that they constitute evidence and should be preserved as such. Powder diffraction is a non-destructive technique and is therefore well suited for forensic analysis. It is also versatile and can be used for the analysis of organic, inorganic, and metallic samples, as well as for the qualitative and quantitative analysis of mixtures of these materials (Higashi et al., 2010; Trzybiński et al., 2013). Geological applications of XRD represent a key tool in mineral exploration. Mineralogists were among the first to develop and promote the new field of X-ray crystallography after its discovery, and so the advent of XRD has literally revolutionized the geological sciences to such an extent that they are unthinkable without this tool. Today, any geological group actively engaged in mineralogical studies would be lost without XRD to clearly characterize individual crystal structures (Shrivastava & America, 2009). Since the microelectronics industry uses silicon and gallium arsenide single crystal substrates in the fabrication of integrated circuits, these materials need to be extensively characterized using XRD. XRD topography can readily detect and image the presence of defects within a crystal, making it a powerful nondestructive evaluation tool for characterizing industrially important single crystal samples. The development of strained semiconductor materials represents an important aspect of improving complementary metal-oxide-semiconductor (CMOS) performance required for current and future generations of microelectronics technology. Understanding the mechanical response of silicon (Si) channel regions and their surroundings is key to predicting and designing device operation. Due to the complexity of the composite, geometries associated with microelectronic are not constant for a given element and show some variation, e.g., chemical environment, glass industry requirements (Murray et al., 2010, 2012). Although glasses are X-ray amorphous and do not themselves provide XRD patterns, there are nevertheless many applications of XRD in the glass industry. These include the identification of crystalline particles that cause minute defects in glass, and measurements of crystalline coatings in terms of texture, crystallite size, and crystallinity. The effects of high pressures on the structure of fused silica have been studied using high-energy XRD up to 43.5 GPa (Benmore et al., 2010). Corrosion analysis XRD is the only analytical method that provides information about the phase composition of solid materials. The most important, versatile, and widely used method for corrosion protection of steel works includes paints or organic coatings. Information on the microscopy of a protective coating is essential to understanding the basic determinants of its properties and improvement requirements. Bitumen has been an important material for the protection of steel mills in the petroleum or other chemical and water industries of the world. However, bitumen has some undesirable properties, and its quality can vary widely from one source to another (Guma et al., 2012).

2.7 SUMMARY

XRD is an analytical technique used to characterize crystalline phases of a variety of materials, usually for mineralogical analysis and identification of unknown substances. Powder diffraction data are basically derived from atomic and molecular arrangements explained by the physics of crystallography. There are several advantages of XRD techniques in scientific laboratories, such as non-destructive, fast, easy sample preparation in XRD, high accuracy for d-spacing calculations, can be performed in situ, permit single crystal, poly, and amorphous materials, and standards are available for thousands of material systems. In recent years, powder XRD systems are becoming more efficient for the pharmaceutical industry due to innovations or improvements in detection and/or source emission technology. XRD methods are particularly important for the analysis of solids in forensics. They are often the only methods that allow further differentiation of materials under laboratory conditions.

REFERENCES

Aaltonen, J., Allesø, M., Mirza, S., Koradia, V., Gordon, K. C., & Rantanen, J. (2009). Solid form screening–A review. In *European Journal of Pharmaceutics and Biopharmaceutics* 71(1), 23–37. https://doi.org/10.1016/j.ejpb.2008.07.014

Andreeva, P., Stoilov, V., & Petrov, O. (2011). Application of X-Ray diffraction analysis for sedimentological investigation of Middle Devonian dolomites from Northeastern Bulgaria. *Geologica Balcanica*, 40(1–3), 31–38. DOI:10.52321/GeolBalc.40.1-3.31

Benmore, C. J., Soignard, E., Amin, S. A., Guthrie, M., Shastri, S. D., Lee, P. L., & Yarger, J. L. (2010). Structural and topological changes in silica glass at pressure. *Physical Review B–Condensed Matter and Materials Physics*, 81(5). https://doi.org/10.1103/PhysRevB.81.054105

Bish D. L. and Post, J. E. (1994). Modern powder diffraction. Reviews in mineralogy, *Acta Cryst*, 20, xi.

Borghetti, G. S., Carini, J. P., Honorato, S. B., Ayala, A. P., Moreira, J. C. F., & Bassani, V. L. (2012). Physicochemical properties and thermal stability of quercetin hydrates in the solid state. *Thermochimica Acta*, 539, 109–114. https://doi.org/10.1016/j.tca.2012.04.015

Brindley, G. W. & Brown, G. (1981). Crystal structures of clay minerals and their X-ray identification. Mineralogical Society. *Clay Minerals 16*, 217–219.

Florian, P., Sadiki, N., Massiot, D., & Coutures, J. P. (2007). 27Al NMR study of the structure of lanthanum–And yttrium-based aluminosilicate glasses and melts. *Journal of Physical Chemistry B*, 111(33), 9747–9757. https://doi.org/10.1021/jp072061q

Guma, T. N., Madakson, P.B., Yawas, D.S., Aku, S.Y. (2012). X-Ray diffraction analysis of the microscopies of some corrosion-protective bitumen coatings. *International Journal of Modern Engineering Research 2*(6), 4387–4395.

Hanawalt, J. D., Rinn, H. W., & Frevel, L. K. (1938). Chemical analysis by X-ray diffraction classification and use of X-ray diffraction patterns. *Industrial & Engineering Chemistry Analytical Edition 10*(9), 457–512, DOI: 10.1021/ac50125a001

Higashi, Y., Smith, T. J., Jez, J. M., & Kutchan, T. M. (2010). Crystallization and preliminary X-ray diffraction analysis of salutaridine reductase from the opium poppy *Papaver somniferum. Acta Crystallographica Section F: Structural Biology and Crystallization Communications, 66*(2), 163–166. https://doi.org/10.1107/S174430910904932X

Ivanisevic, I., McClurg, R. B., & Schields SSCI, P. J. (2010). Uses of X-Ray powder diffraction in the pharmaceutical industry. In *Pharmacetical Science Encyclopedia: Drug Discovery, Development and Manufacturing*, edited by S. C. Gad, pp. 1–42. New York: John Wiley & Sons, Inc.

Jackson, M. L. (1981). An X-ray diffraction refinement of the structure of natural natrolite. *Acta Crystallographica Section B: Structural Crystallography and Crystal Chemistry, 37*(10), 1909–1911.

Junhui, L., Ruishan, W., Lei, H., Fuliang, W., & Zhili, L. (2011). HRTEM and X-ray diffraction analysis of Au wire bonding interface in microelectronics packaging. *Solid State Sciences, 13*(1), 72–76. https://doi.org/10.1016/j.solidstatesciences.2010.10.011

Klug, H. P., & Alexander, L. E. (1974). *X-Ray Diffraction Procedures for Polycrystalline and Amorphous Materials*. New York: John Wiley & Sons, 960.

Mendenhall, M. H., Mullen, K., & Cline, J. P. (2015). An implementation of the fundamental parameters approach for analysis of X-ray powder diffraction line profiles. *Journal of Research of the National Institute of Standards and Technology, 120*, 223–251. https://doi.org/10.6028/jres.120.014

Moore, D. M. & Reynolds, R. C. (1989). *X-Ray Diffraction and the Identification and Analysis of Clay Minerals*, New York: Oxford University Press, 322.

Murray, C. E., Ryan, E. T., Besser, P. R., Witt, C., Jordan-Sweet, J. L., & Toney, M. F. (2012). Understanding stress gradients in microelectronic metallization. *Powder Diffraction, 27*(2), 92–98. https://doi.org/10.1017/S0885715612000231

Murray, C. E., Ying, A. J., Polvino, S. M., Noyan, I. C., & Cai, Z. (2010). Nanoscale strain characterization in microelectronic materials using X-ray diffraction. *Powder Diffraction, 25*(2), 108–113. https://doi.org/10.1154/1.3394205

Newman, A. (n.d.). *X-ray Powder Diffraction in Solid Form Screening and Selection XRD » X-ray Powder Diffraction in Solid Form Screening and Selection.* www.researchgate.net/publication/230651007

Palmieri, M., Vagnini, M., Pitzurra, L., Rocchi, P., Brunetti, B. G., Sgamellotti, A., & Cartechini, L. (2011). Development of an analytical protocol for a fast, sensitive and specific protein recognition in paintings by enzyme-linked immunosorbent assay (ELISA). *Analytical and Bioanalytical Chemistry, 399*(9), 3011–3023. https://doi.org/10.1007/s00216-010-4308-1

Shrivastava, V. S., & America, S. (2009). X-ray diffraction and mineralogical study of soil: A review. *Journal of Applied Chemical Research, 9*, 41–51. www.SID.ir

Trzybiński, D., Niedziałkowski, P., Ossowski, T., Trynda, A., & Sikorski, A. (2013). Single-crystal X-ray diffraction analysis of designer drugs: Hydrochlorides of metaphedrone and pentedrone. *Forensic Science International, 232*(1–3). https://doi.org/10.1016/j.forsciint.2013.07.012

Vogt, F. G., & Williams, G. R. (2011). ChemInform abstract: Advanced approaches to effective solid-state analysis: X-ray diffraction, vibrational spectroscopy and solid-state NMR. *ChemInform, 42*(18), 58–65. https://doi.org/10.1002/chin.201118276

Waseda, Y., Matsubara, E., & Shinoda, K. (n.d.). *X-Ray Diffraction Crystallography X-Ray Diffraction Crystallography Introduction, Examples and Solved Problems.* Springer, 67–106, doi:10.1007/978-3-642-16635-8_3

3 Nanomechanical System

A nanoindenter is the main component for indentation hardness tests used in nanomechanical systems. Since the mid-1970s, the nanomechanical system has become the main method for measuring and testing very small amounts of mechanical properties. The nanomechanical system also referred to as "depth sensing indentation" (DSI) or "instrumented indentation," gained popularity with the development of machines that can record small loads and displacements with high accuracy and precision (Randall et al., 2009; Xiao et al., 2021) The load-displacement data can be used to determine the modulus of elasticity, hardness, yield strength, fracture toughness, scratch hardness, and wear properties (Oyen & Cook, 2009).

There are many types of nanoindenters currently in use, distinguished mainly by their tip geometry. Among the geometry available are three and four-sided pyramids, wedges, cones, cylinders, filaments, and spheres. Some geometries have become standard due to their widespread use and known properties, such as Berkovich, cube-corner, Vickers, and Knoop nanoindenters. To meet the high precision requirements, nanoindenters must be manufactured according to the definitions of ISO 14577-2 and tested and measured with equipment and standards traceable to the National Institute of Standards and Technology (NIST). The tip of the indenter may be sharp, flat, cylindrical, or spherically rounded. The material for most nanoindenters is diamond and sapphire, although other hard materials such as quartz, silicon, tungsten, steel, tungsten carbide, and almost any other hard metal or ceramic material can be used. Diamond is the most commonly used material for nanomechanical systems because of its hardness, thermal conductivity, and chemical inertness. In some cases, the electrically conductive diamond may be required for special applications and is also available. Nanoindenters are mounted on holders, which may be the standard design of a manufacturer of nanoindenter devices or custom-made. The material of the holder can be steel, titanium, machinable ceramics, other metals, or rigid materials. In most cases, the indenter is attached to the holder using a rigid metal bonding process. The metal forms a molecular bond with both materials, be it diamond steel, diamond ceramic, etc.

The dimensions of nanoindenters are very small, some less than 50 microns, and they are manufactured with precise angular geometry to achieve the highly accurate readings required for nanomechanical systems. Instruments for measuring

DOI: 10.1201/9781003340546-4

angles on larger objects, such as protractors or comparators, are not practical or precise enough for measuring nanoindenter angles, even using microscopes. For precise measurements, a laser goniometer is used to measure the angles of diamond nanoindenters. The nanoindenter surfaces are highly polished and reflective, which is the basis for the laser goniometer measurements. The laser goniometer can measure specific or desired angles with an accuracy of one-thousandth of a degree (Winer et al., n.d.).

3.1 CONSTRUCTION

Nanoindenter instruments are used to measure the mechanical properties of materials. The nanomechanical system provides data for mechanical characterization of surfaces by mechanically scanning a material surface to a depth in the nanometer range. Mechanical properties measured include hardness, elastic-plastic deformation, Young's modulus, film-substrate scratch resistance, film-substrate adhesion, film residual stress, time-dependent creep film, relaxation properties, fracture toughness film, and fatigue. In this research, roughness, hardness, modulus, and stiffness of the thin film coatings are measured using a nanoindenter (Ling, n.d.).

The nanoindenter consists of a coil and magnet assembly located at the top of the loading column that is used to move the indenter toward the sample as shown in Figure 3.1(a). The load generated is the vector product of the magnetic field strength of the permanent magnet and the current load through the coil. This type of load application is very simple to use and very linear in calibration. It also allows very fast feedback control of the displacement because it completely separates the load application systems and the displacement measurement systems (Loubet, 1984, n.d.). The depth of the indenter is measured with a capacitive displacement meter. The nanoindenter is equipped with a device for measuring stiffness. In addition to this component, the other important devices are the indenter head, an optical/grating probe microscope, and an x-y-z motorized precision stage to position and transport the sample

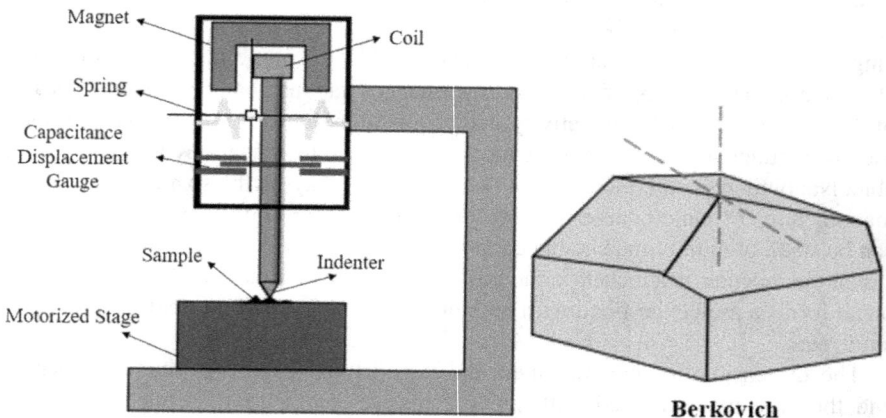

FIGURE 3.1 Nanoindenter: (a) Apparatus and (b) Berkovich Tip.

for indentation. The most commonly used tip for nanomechanical systems is the Berkovich tip (Oliver & Pharr, 2004) shown in Figure 3.1(b). The total incorporated angle at this tip is 142.3°. This tip geometry is the standard for nanomechanical systems because it can be easily shaped into a sharper tip.

The indentation technique originated with Moh's scale of hardness in 1822, in which materials capable of leaving a permanent scratch in another material were classified as harder, with diamond assigned the maximum value of 10 on the scale. The establishment of the Brinell, Knoop, Vickers, and Rockwell tests is based on a refinement of the method of indenting one material with another. The nanomechanical system is simply an indentation test in which the length scale of indentation is measured in nanometers rather than micrometers or millimeters, as is common in conventional hardness tests. Apart from the displacement scale, the distinguishing feature of most nanomechanical testing systems is the indirect measurement of the contact area, i.e., the contact area between the indenter and the specimen. In conventional indentation tests, the contact area is calculated from direct measurements of the dimensions of the residual indentation that remains on the specimen surface after the load is removed. For tests using nanomechanical systems, the size of the residual indentation is in the order of micrometers and is too small to be measured directly. Therefore, it is common to determine the contact area by measuring the depth of penetration of the indenter into the specimen surface. Combined with the known geometry of the indenter, this allows indirect measurement of the contact area at full load. For this reason, the testing of nanomechanical systems is sometimes referred to as DSI.

Hardness is not the only parameter of interest to materials scientists. Indentation methods can also be used to calculate the modulus of elasticity, strain hardening exponent, fracture toughness (for brittle materials), and viscoelastic properties. How can such a variety of properties be extracted from such a simple test, which in many ways can be considered a "non-destructive" test method? Consider the force-displacement behavior shown in Figure 3.3. This type of data is obtained when a spherical indenter is placed in contact with the flat surface of the specimen with a steadily increasing force. Both load and indentation depth are recorded at each loading step (ultimately providing a measure of modulus and hardness as a function of depth below the surface). After the maximum load is reached, for the material shown in Figure 3.2(a), the

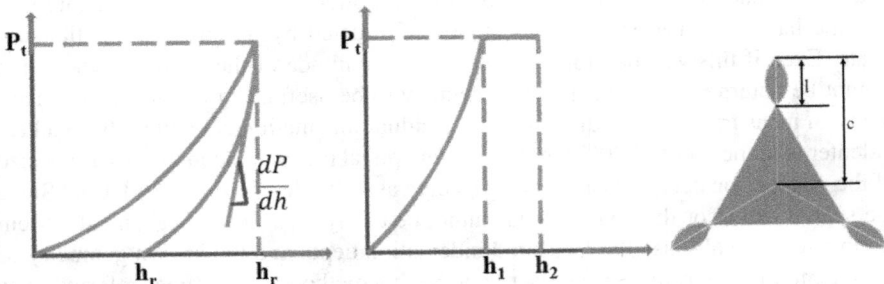

FIGURE 3.2 Load-displacement curves for (a) an elastic plastic solid and (b) a viscoelastic solid for a spherical indenter and (c) cracks emanating from the corners of the residual impression in a brittle material.

load is steadily reduced and the penetration depth is recorded. The loading portion of the indentation cycle may consist of initial elastic contact followed by plastic flow or flow within the specimen at higher loads. During unloading, if the flow has occurred, the force-displacement data follows a different path until a residual indentation is left in the specimen surface at zero loads. The maximum depth of indentation for a given load, together with the slope of the unloading curve measured at the tangent to the data point at maximum load, results in a measure of both the hardness and elastic modulus of the specimen material. In some cases, it is possible to measure the modulus of elasticity not only from the unloading section but also from the loading section of the curve (Bernachy-Barbe, 2019). For a viscoelastic material, the relationship between load and penetration is not so simple. For such materials, the indentation test is accompanied by "creep," which is manifested by a change in indentation depth at constant load, as shown in Figure 3.2(b). Analysis of the creep component of the load-displacement response provides quantitative information about the elastic "solid-like" properties of the specimen and also about the "fluid-like" or "out-of-phase" components of the specimen properties. In brittle materials, cracking of the specimen may occur, especially when using a pyramid-shaped indenter such as the three-sided Berkovich or the four-sided Vickers indenter. As shown in Figure 3.2(c), the length of the crack, which often starts at the corners of the indentation, can be used to calculate the fracture toughness of the specimen material. More advanced methods can be used to study residual stresses in thin films, the properties of materials at high temperatures, scratch resistance and film adhesion, and in some cases Van der Waals surface forces. This book examines and reports on all of these topics, beginning with a description of the test method and the basis on which the analysis is based. Later chapters deal with the various corrections required to account for a number of instrumental and material effects that are a source of error in measurement, the theoretical aspects behind the constitutive laws relating mechanical properties to measured quantities, recent attempts to formulate an international standard for nanomechanical systems, examples of applications, and a brief description of commercially available instruments.

The current field of nanomechanics grew out of a desire to measure the mechanical properties of hard thin films and other near-surface treatments in the early 1980s. The microhardness testers available at the time could not apply sufficiently low forces to achieve penetration depths of less than the required 10% of the film thickness so that the hardness measurement would not be affected by the presence of the substrate. Even if this was possible, the resulting magnitude of the residual indentation cannot be determined with sufficient accuracy to be useful. For example, the uncertainty in measuring a 5 μm diagonal of a residual indentation created with a Vickers indenter is in the order of 20% when using an optical method and increases as the size of the indentation decreases and can be as high as 100% for a 1 μm indentation. Since the spatial extent of the contact area cannot be easily measured, mechanical system techniques typically use the measured indentation depth and the known geometry of the indenter to determine the contact area. Such a method is sometimes referred to as "depth of indentation measurement." In order for such a measurement to be made, the depth measurement system must be referenced to the sample surface. This is usually done by bringing the indenter into contact with the surface with a very low "initial contact force," which in turn results in an inevitable initial indentation of the indenter into the surface that must be accounted for in the analysis (Bendavid et al., 2000). Additional corrections are required to account for irregularities in the shape of the indenter, deflection of the loading frame, and piling of material around the indenter (see Figure 3.3). These effects contribute to errors in the recorded depths and subsequently in the determination of hardness and modulus. Furthermore, the scale of deformation in a nanomechanical system test becomes comparable to the magnitude of material defects such as dislocations and grain sizes, and the continuum approximation used in the analysis may lose validity.

The test results of the nanomechanical system provide information on elastic modulus, hardness, strain hardening, cracking, phase transformation, creep, and energy absorption. Specimen size is very small and testing can be considered nondestructive in many cases. Specimen preparation is straightforward. Since the extent of deformation is very small, the technique can be applied to thin surface films and surface-modified coatings. In many cases, the microstructural features of a thin film or coating differ significantly from those of the base material due to residual stresses, preferred orientations of crystallographic planes, and microstructure morphology. Applications of the technique, therefore, include technologies such as cathodic arc deposition, physical vapor deposition (PVD), and chemical vapor deposition (CVD), as well as ion implantation and functionally graded materials. Instruments for nanomechanical systems are typically easy to use, operate under computer control, and do not require vacuum chambers or other expensive laboratory infrastructure (Chen et al., 2015). The technique is based on continuous measurement of penetration depth with increasing load and was apparently first demonstrated by Pethica and Oliver (1982) and has been used to measure the mechanical properties of ion-implanted metal surfaces, a popular application of this technique for many years. The idea of using elastic recovery of hardness indentations to determine mechanical properties is not new (Lawn & Howes, 1981). Today's modern treatments probably start with how to measure contact area using the unloading portion of the load-displacement curve (Loubet 1984, n.d.). This method is used for relatively high loads (in the order of 1 newton) (Bor et al., 2019). The most commonly used method of

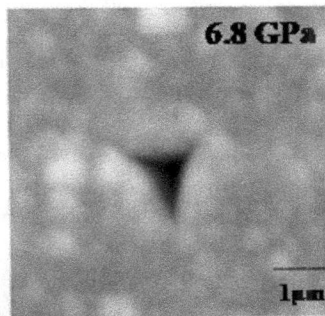

FIGURE 3.3   Atomic force micrograph of a residual impression in steel made with a triangular pyramid Berkovich indenter.

analysis is a refinement of Doerner and Nix's approach , which was subsequently shown to be equivalent to the (Oliver & Pharr, 1992). The first "ultramicro" hardness tests were performed using a device designed for use in the vacuum chamber of a scanning electron microscope (SEM), in which a sharp-edged tungsten wire was loaded by the movement of a galvanometer controlled externally by electric current. The depth of penetration was determined by measuring the motion of the penetrator using an interferometric method. The later use of strain gauges to measure the applied force and of finely machined parallel springs actuated by an electromagnetic coil brought measurement outside the vacuum chamber into the laboratory, but although the required forces could now be applied in a controlled manner, the optical measurement of displacement or the magnitude of residual indentations remained a limiting factor. Developments in electronics led to the production of sensors for displacement measurement that offered higher resolution than optical methods, and in the last ten years some six or seven instruments have become commercial products, often leading to the formation of private companies spun out of research institutions to sell and support them (Bell et al., 1991). There is no doubt that as mechanisms become smaller, interest in mechanical properties at the nanometer scale and below, and in the nature of surface forces and adhesion, will continue to grow. Indeed, in at least one recent publication, the combination of a nanoindenter and an atomic force microscope is referred to as a "picoindenter" (Syed Asif et al., 2000) suitable for the study of pre-contact mechanics, the process of contact initiation, and the actual contact mechanics. The current maturity of the field of nanomechanics makes it a suitable technique for the evaluation of new materials technologies by academic and private sector research laboratories, and it is finding increasing application as a quality control tool.

3.2 WORKING PRINCIPLE

The nanomechanical testing system is used to measure hardness and elastic modulus by instrumented indentation techniques and is used to characterize the mechanical behavior of materials on a small scale. Its attractiveness is mainly based on the fact that mechanical properties can be determined directly from indentation force and displacement measurements without the need to image the hardness indentation, which enables the measurement of properties in the micrometer and nanometer range with high-resolution testing equipment. For this reason, the method has become a primary technique for determining the mechanical properties of thin films and small structural features. The nanomechanical test system of Hysitron make is shown in Figure 3.4(a). The method was developed to measure the hardness and elastic modulus of material from indentation force-displacement data obtained during a loading and unloading cycle, and Figure 3.4(b) and (c) show the schematic representation of the indentation characteristics. Although the method was originally intended for use with sharp, geometrically self-similar indenters, such as the Berkovich triangular pyramid, we have since realized that it is much more general and applicable to a variety of axisymmetric indenters geometries including the sphere. Hardness and Young's modulus are determined using the analytical method of Oliver and Pharr.

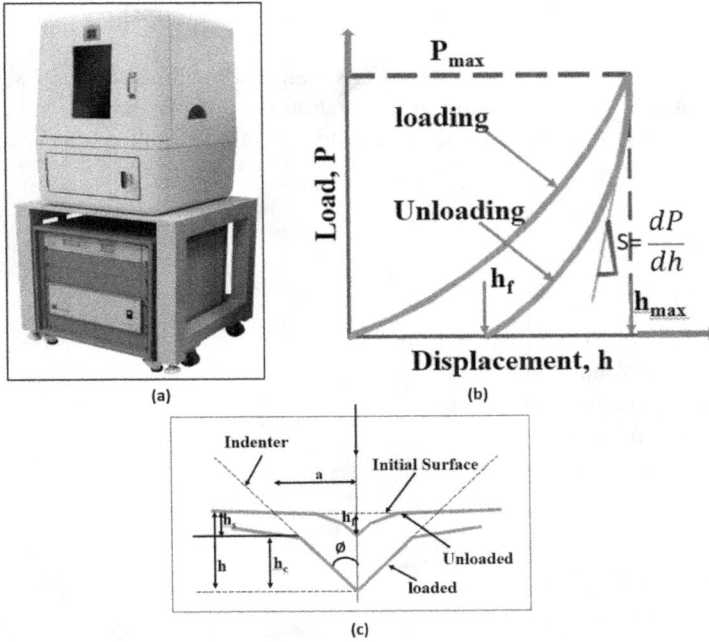

FIGURE 3.4 Schematic illustration of (a) figure of nanomechanical test system, (b) indentation load-displacement data, and (c) unloading process showing parameters characterizing the contact geometry.

However, to avoid calculating the contact area, which sometimes leads to errors in the results due to sinking or buildup, the method of Oliver and Pharr (2004) is used. The hardness of a nanomechanical system is defined as the indentation force (P) divided by the projected contact area (A) of the indentation. It is the average pressure that a material can withstand under load. From the load-displacement curve, the hardness at peak load can be derived as follows:

$$H = \frac{P_{max}}{A} \qquad (3.1)$$

The modulus of elasticity of the compressed specimen can be derived from the initial unloading contact stiffness,

$$S = \frac{dP}{dh} \qquad (3.2)$$

Figure 3.4(b) shows the typical force-displacement curve and deformation pattern of an elastic-plastic specimen during and after indentation. In Figure 3.4(c), h_{max} represents the displacement at the peak load P_{max}. h_c is the contact depth and is defined as the depth of the indenter in contact with the specimen under load. hf is the final displacement after complete unloading. S is the initial contact stiffness at unloading.

If Young's modulus E_f of the film differs from that of the substrate Es, then the measured value E_r changes with the indentation depth h. If the film has a similar Young's modulus to the substrate, then the reduced Young's modulus can be expressed as follows:

$$\frac{1}{E_r} = \frac{1-v_i^2}{E_i} + \frac{1-v_f^2}{E_f}$$

(3.3)

where,

E_r – Reduced Young's modulus
E_i – Young's modulus of indenter
E_f – Young's modulus of film
v_i – Poisson's ratio of indenter
v_f – Poisson's ratio of the film

3.3 APPLICATIONS

The nanomechanical system is a method for measuring the mechanical properties of materials in small volumes using an instrumented indentation technique. Modulus of elasticity, hardness, fracture toughness, creep behavior, and dynamic properties such as storage and loss modulus can be measured. Nanomechanical systems can be used for a wide range of applications. Some applications require precise XY positioning of the indentation. An example is the measurement of the mechanical properties of coke as used in the mineral processing industry (Syed Asif et al., 2000). In this application, indentations must be placed in the center of the particles so that the readings are not affected by the surrounding matrix. An optical microscope is used to position the indentation before testing and to view the impression after testing. Typically, an XY positioning accuracy of about 0.5 μm is required. Another common application of nanomechanical systems is thin film testing (Lupinacci et al., 2015). TiN coatings are often used as hard coatings on cutting tools. Manufacturers of such coatings are interested in how the hardness of the film changes under different machining conditions. These coatings are usually quite rough and have many irregularities and grain boundaries. To determine the hardness of the coating, it is usually necessary to perform a series of tests with increasing maximum load. At low loads, the values of H are influenced by the developing plastic zone (Deng et al., 2021). At a high load, the results are influenced by the presence of the substrate. The film hardness is usually taken to be that in a plateau region intermediate between the two. Scratch testing is a popular application of the nanomechanical system and is the related test of scratch testing on a micron scale. Depending on the instrument's capabilities, scratch testing can be done by moving the sample relative to the indenter tip while the load is applied (either ramped or steady) to the indenter (Field & Swain, n.d.). At the same time, the lateral force required to move the specimen is usually measured. The ratio of the lateral force to the normal force is the coefficient of friction at the contact (Karimzadeh et al., 2019). Other applications include multiphase steels, thermal barrier coatings, biological samples such as teeth and bones, soft films (paint), heat-treated surfaces, and semiconductor wafers.

3.4 SUMMARY

In the last two decades, there has been considerable interest in the mechanical characterization of thin films and small volumes of materials by means of depth-sensitive indentation tests using indenters. Typically, the main objective of such tests is to obtain values for the modulus of elasticity and hardness of the test material from experimental measurements of indentation force and depth (Wang et al., 2018). The forces involved are usually in the millinewton range and are measured with a resolution of a few nano-newtons. The penetration depths are in the order of nanometers, hence the name "nanomechanical system." This technique is one of the most widely used characterization methods for determining mechanical properties in quality control and failure analysis research.

REFERENCES

Bell, T. J., Bendeli, A., Field, J. S., Swain, M. v., & Thwaite, E. G. (1991). The determination of surface plastic and elastic properties by ultra micro-indentation. *Metrologia*, *28*. http://iopscience.iop.org/0026-1394/28/6/004

Bendavid, A., Martin, P. J., & Takikawa, H. (2000) Deposition and modification of titanium dioxide thin films by filtered arc deposition. *Thin Solid Films*, *360*, 241–249, DOI:10.1016/S0040-6090(99)00937-2

Bernachy-Barbe, F. (2019). A data analysis procedure for phase identification in nanoindentation results of cementitious materials. *Materials and Structures/Materiaux et Constructions*, *52*(5). https://doi.org/10.1617/s11527-019-1397-y

Bor, B., Giuntini, D., Domènech, B., Swain, M. v., & Schneider, G. A. (2019). Nanoindentation-based study of the mechanical behavior of bulk supercrystalline ceramic-organic nanocomposites. *Journal of the European Ceramic Society*, *39*(10), 3247–3256. https://doi.org/10.1016/j.jeurceramsoc.2019.03.053

Chen, L., Ståhl, J. E., & Zhou, J. (2015). Analysis of in situ mechanical properties of phases in high-alloyed white iron measured by grid nanoindentation. *Journal of Materials Engineering and Performance*, *24*(10), 4022–4031. https://doi.org/10.1007/s11665-015-1672-1

Deng, B., Luo, J., Harris, J. T., Smith, C. M., & Wilkinson, T. M. (2021). Toward revealing full atomic picture of nanoindentation deformation mechanisms in Li_2O-$2SiO_2$ glass-ceramics. *Acta Materialia*, *208*. https://doi.org/10.1016/j.actamat.2021.116715

Field, J. S., & Swain, M. v. (n.d.). (1993). A simple predictive model for spherical indentation. *J. Mater. Res.*, *2*(8), 297–306.

Karimzadeh, A., Koloor, S. S. R., Ayatollahi, M. R., Bushroa, A. R., & Yahya, M. Y. (2019). Assessment of nano-indentation method in mechanical characterization of heterogeneous nanocomposite materials using experimental and computational approaches. *Scientific Reports*, *9*(1). https://doi.org/10.1038/s41598-019-51904-4

Lawn, B. R., & Howes, V. R. (1981). Elastic recovery at hardness indentations. *Journal of Materials Science*, *16*, 2745–2752.

Loubet, J. L., Georges, J. M. Marchesini, O., Meille, G. (1984). Vickers indentation curves of magnesium oxide (MgO), *Journal of Tribology*, *106*, 43–48.

Lupinacci, A., Chen, K., Li, Y., Kunz, M., Jiao, Z., Was, G. S., Abad, M. D., Minor, A. M., & Hosemann, P. (2015). Characterization of ion beam irradiated 304 stainless steel utilizing nanoindentation and Laue microdiffraction. *Journal of Nuclear Materials, 458*, 70–76. https://doi.org/10.1016/j.jnucmat.2014.11.050

Oliver, W. C., & Pharr, G. M. (1992). An improved technique for determining hardness and elastic modulus using load and displacement sensing indentation experiments. *Journal of Materials Research, 7*(6), 1564–1583. https://doi.org/10.1557/jmr.1992.1564

Oliver, W. C., & Pharr, G. M. (2004). Measurement of hardness and elastic modulus by instrumented indentation: Advances in understanding and refinements to methodology. *Journal of Materials Research, 19*(1), 3–20, https://doi.org/10.1557/jmr.2004.19.1.3

Oyen, M. L., & Cook, R. F. (2009). A practical guide for analysis of nanoindentation data. *Journal of the Mechanical Behavior of Biomedical Materials, 2*(4), 396–407. https://doi.org/10.1016/j.jmbbm.2008.10.002

Pethica, J. B., & Oliver, W. C. (1982). *Ultra-MIcrohardness Tests on Ion Implanted Metal Surfaces.* Elsevier Science Publishing Company, 372379.

Randall, N. X., Vandamme, M., & Ulm, F. J. (2009). Nanoindentation analysis as a two-dimensional tool for mapping the mechanical properties of complex surfaces. *Journal of Materials Research, 24*(3), 679–690. https://doi.org/10.1557/jmr.2009.0149

Syed Asif, S. A., Wahl, K. J., & Colton, R. J. (2000). Quantitative study of nanoscale contact and pre-contact mechanics using force modulation. *Beilstein Journal of Nanotechnology, 10*(1), 1332–1347, DOI:10.3762/bjnano.10.132

Wang, H., Zhu, L., & Xu, B. (2018). Principle and methods of nanoindentation test. In *Residual Stresses and Nanoindentation Testing of Films and Coatings* (pp. 21–36). Springer Singapore. https://doi.org/10.1007/978-981-10-7841-5_2

Winer, W. O., Bergles, A. E., Klutke, G. A., Wang, K. K., Finnie, I., Welty, J. R., Bryant, M. D., Yang, H. T., Mow, V. C., Leckie, F. A., & Gross, D. (n.d.). *Nanoindentation, Mechanical Engineering Series.* www.springer.com/series/1161

Xiao, H., Wang, X., & Long, C. (2021). Theoretical model for determining elastic modulus of ceramic materials by nanoindentation. *Materialia, 17.* https://doi.org/10.1016/j.mtla.2021.101121

4 X-Ray Photo Spectroscopy (XPS)

X-ray photoelectron spectroscopy (XPS) is one of the most commonly used surface-sensitive techniques. It is a surface-sensitive analytical technique in which X-rays bombard the surface of a material and the kinetic energy of the emitted electrons is measured. The two key features of this technique that make it a powerful analytical method are its surface sensitivity and its ability to determine information about the chemical state of elements in the sample. All elements except hydrogen and helium can be detected. The method has been used to study the surface of almost any material, from plastics to textiles to soil to semiconductors. All materials have surfaces, and these surfaces interact with other materials. Factors such as surface wettability, adhesion, corrosion, charge transfer, and catalysis are all governed by surfaces and surface impurities, making the study and understanding of surfaces important. XPS is based on the photoelectric effect, first discovered by Heinrich Hertz in 1887. He found that electrons are emitted from surfaces when they are irradiated with light. This chapter, therefore, explains in detail how XPS works, how it is constructed, and how it is used.

4.1 CONSTRUCTION

XPS belongs to the family of photoemission spectroscopy, in which the large electron spectra are obtained by irradiating a material with an X-ray beam. Chemical states are derived from the measurement of kinetic energy (KE) and the number of electrons knocked out. XPS requires high vacuum (residual gas pressure $p \sim 10^{-6}$ Pa) or ultrahigh vacuum conditions ($p < 10^{-7}$ Pa), although a current area of development is ambient pressure XPS, in which samples are analyzed at pressures of a few tens of millibars. When laboratory X-ray sources are used, XPS can readily detect all elements except hydrogen (H_2) and helium (He). XPS is routinely used for the analysis of inorganic compounds, semiconductors, catalysts, alloys, polymers, glasses, paints, makeup, teeth, bones, implants, biomaterials, viscous oils, ion-modified materials, etc. (Gumerova & Rompel, 2018; Li et al., 2018; Rahmayeni et al., 2019) To some extent and less routinely, XPS is used to analyze hydrated forms of materials such as hydrogels and biological samples by freezing them in their hydrated state in an ultra-clean environment and allowing several layers of ice to sublime before analysis.

DOI: 10.1201/9781003340546-5

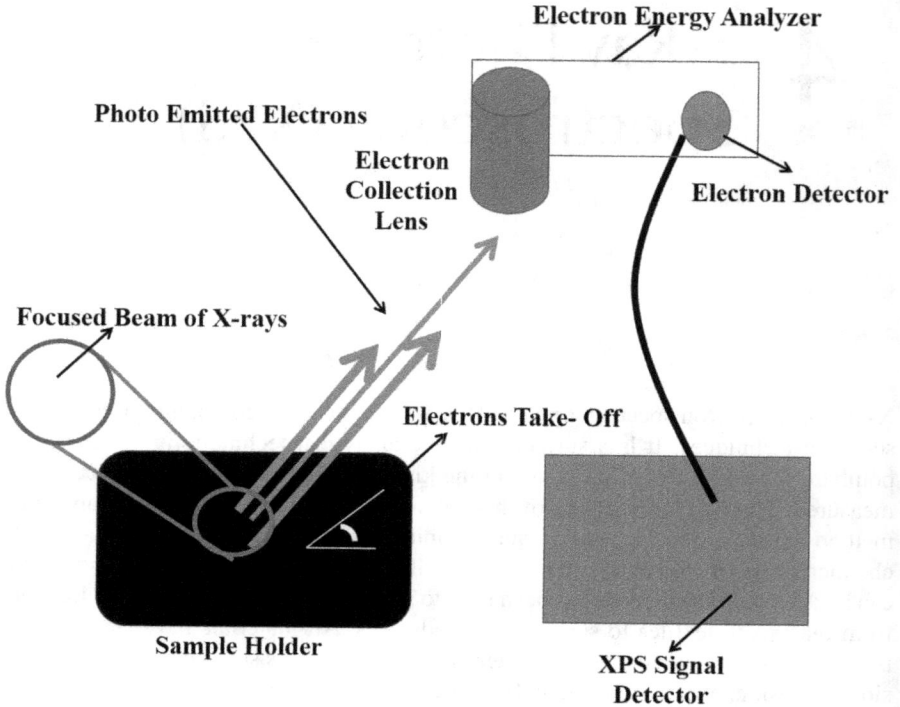

FIGURE 4.1   Basic components of an XPS system.

$$E_{binding} = E_{photon} - (E_{kinetic} + \phi)$$

Here, the binding energy is the energy of an electron attracted to a nucleus; the photon energy is the energy of the X-ray photons used by the spectrometer, and the kinetic energy is the energy of the electrons ejected from the sample. The work function is a correction factor for the instrument and is the minimum energy required to eject an electron from an atom (more information on the photoelectric effect, but not required for understanding). The work function and photon energy are known, and the kinetic energy is measured by the detector. This leaves the binding energy as the only unknown, which can then be determined. The farther the electrons in the orbitals are from the nucleus, the less energy is required to eject them, so the binding energy is lower at higher orbitals. Electrons in different subshells (s, p, d, etc.) also have different energies. By displaying the energy of the electrons emitted from a material, XPS can be used to determine the composition of a material (see Figure 4.1) (Naumann D'Alnoncourt et al., 2014).

4.2   WORKING PRINCIPLE

X-rays of a specific wavelength are selected based on the material being tested and the kinetic energies of the emitted electrons are measured; the binding energy of individual electrons can be determined using the photoelectric equation (Huang & Zhang,

2022). Based on Einstein's photoelectric effect, many materials emit electrons when irradiated with light (Figure 4.2).

$$E_K = h\nu - E_b - \varphi_{sp}$$

E_K = kinetic energy
E_b = binding energy
h = Planck's constant
ν = frequency of X-rays
φ_{sp} = spectrometer work function

Each atom has a unique XPS spectrum; XPS can be used to determine the elemental composition, stoichiometry, and electrical/chemical states, and to study surface impurities. Figure 4.3 shows the photoelectric effect pertaining to XPS, XPS is an

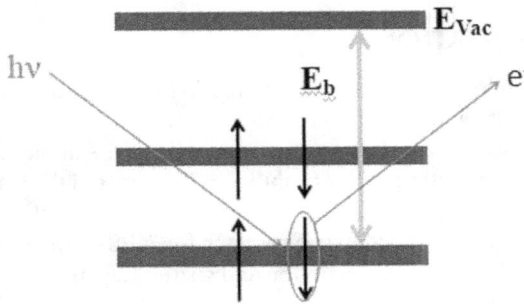

FIGURE 4.2 Photoelectric effect.

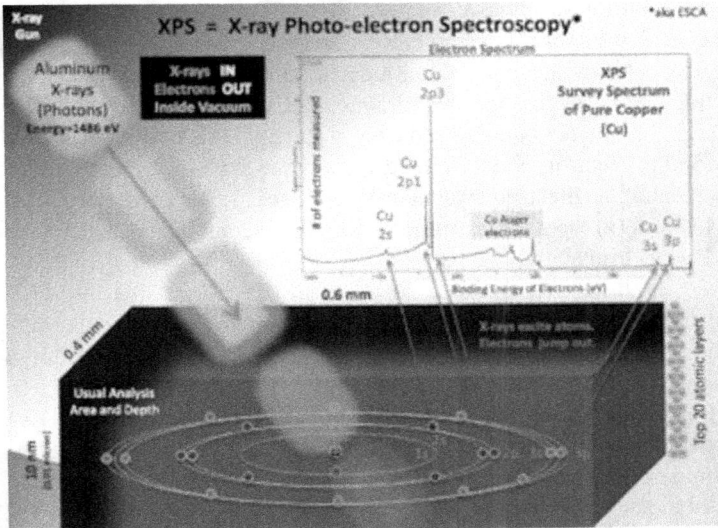

FIGURE 4.3 Photoelectric effect pertaining to XPS. (https://commons.wikimedia.org/wiki/File:XPS_PHYSICS.png).

elemental analysis technique that uniquely provides information about the chemical state of detected elements, such as the distinction between sulfate and sulfide forms of the element sulfur, the binding energy (BE) of the electron measured relative to the chemical potential, the photon-electron is the energy of the X-ray photons used, the kinetic energy of the electron as measured by the instrument, the work function is in terms of the specific surface area of the material, which in real measurements includes a small correction by the instrument work function due to the contact potential. This equation is essentially conservation of energy equation (Voiry et al., 2018). The term similar to the work function can be viewed as an adjustable instrumental correction factor that accounts for the few eV of kinetic energy given off by the photoelectron as it is emitted from the mass and absorbed by the detector. It is a constant that rarely needs to be adjusted in practice. The high-resolution spectra display a quantification indicating the atomic species and their atomic percentages, as well as the characteristic binding energy.

4.3   INSTRUMENTATION

The main components of an XPS system are the X-ray radiation source, an ultra-high vacuum chamber with metal magnetic shielding, an electron compilation lens, an electron energy analyzer, an electron detector system, a sample opening chamber, sample holders, a sample stage with the ability to heat or cool the sample, and a set of stage manipulators (Ray & Shard, 2011).

The most widely used electron spectrometer for XPS is the hemispherical electron analyzer. It has high energy resolution and spatial selection of emitted electrons. However, much simpler electron energy filters – the cylindrical mirror analyzers – are sometimes used, usually to check the elemental composition of the surface. They represent a compromise between the need for high count rates and high angular/energy resolution (Vashishtha & Sapate, 2018). This type consists of two coaxial cylinders placed in front of the sample, with the inner cylinder held at a positive potential while the outer cylinder is held at a negative potential. Only electrons with the correct energy can pass through this arrangement and are detected at the end. Count rates are high, but the resolution (both energy and angle) is poor (Rawat et al., 2020).

An incoming electron is accelerated toward the wall, where it carries off more electrons, creating an electron avalanche until a measurable current pulse is reached. Figure 4.4 shows (a) the XPS instrumentation, (b) vacuum balance, (c) X-ray source, and (d) XPS electron analyzer.

a. Ultra-high vacuum system: allows longer photoelectron transit times, ultra-high vacuum keeps surfaces clean and prevents impurities from producing X-ray signals.

b. XPS measurements are conducted under ultra-high vacuum ($< 10^{-8}$ Torr) in order to avoid collision between photoelectrons and gas molecules in the spectrometer, and minimize surface contamination from residual gases.

c. X-ray source: double anode X-ray source, Mg-Kα radiation: hv = 1253.6 eV, Al-Kα radiation: hv = 1486.6 eV, monochromatic with quartz crystal.

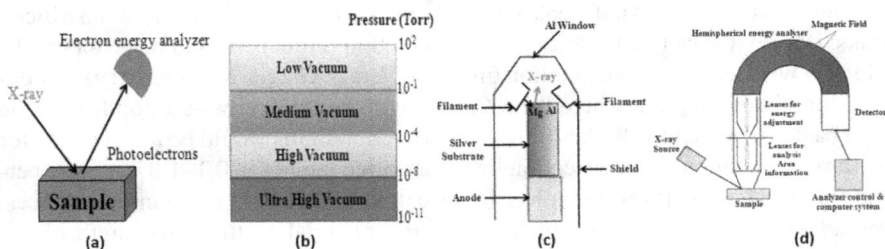

FIGURE 4.4   (a) XPS instrumentation, (b) vacuum scale, (c) X-ray source, and (d) XPS electron analyzer.

d. Electron analyzer: lens system to collect the photoelectrons, analyzer to filter the electron energies, detector to count the electrons.

The typical XPS spectrum is a representation of the number of electrons detected at a given binding energy. Each element produces a series of characteristic XPS peaks. These peaks correspond to the electron configuration of the electrons in the atoms, e.g., 1s, 2s, 2p, 3s, etc. The number of electrons detected in each peak is directly related to the amount of the element in the XPS sampling volume. To determine the percent of atoms, each raw XPS signal is corrected by dividing the intensity by a relative sensitivity factor and normalizing to all detected elements. Since hydrogen is not detected, these atomic percentages exclude hydrogen (Siegbahn & Edvarson, 1958). Quantitative accuracy and precision in XPS are often used to establish an empirical formula because they provide excellent quantitative accuracy for homogeneous solid materials. Absolute quantification requires the use of certified (or independently tested) standard samples and is generally more difficult and less common. Relative quantification involves comparisons between multiple samples in a set where one or more analytes are varied while all other components (the sample matrix) are held constant. Quantitative accuracy depends on several parameters, such as signal-to-noise ratio, peak intensity, the accuracy of relative sensitivity factors, correction for electron transmission function, homogeneity of surface volume, correction for the energy dependence of electron mean free path, and degree of sample degradation due to analysis. Under optimal conditions, the quantitative accuracy of the atomic percentages (at%) calculated from the main XPS peaks is 90–95% for each peak. The quantitative accuracy for the weaker XPS signals, whose peak intensity is 10–20% of the strongest signal, is 60–80% of the true value and depends on the effort made to improve the signal-to-noise ratio (e.g., by signal averaging) (Siegbahn & Edvarson, 1958). Quantitative precision (the ability to repeat measurement and obtain the same result) is an essential factor for proper reporting of quantitative results.

Detection limits can vary widely depending on the cross-section of the effect of the core state of interest and the background signal. In general, the effective cross-section of photoelectrons increases with atomic number. The background increases with the atomic number of the matrix constituents as well as the binding energy due to the secondary emitted electrons. For example, in the case of gold on silicon, where the high cross-section Au4f peak has higher kinetic energy than the main silicon peaks, it is on a very low background and detection limits of 1 ppm or better can be achieved with reasonable acquisition times (de Vera & Garcia-Molina, 2019). In contrast, for silicon on gold, where the Si2p line with its small cross-section lies on the large background below the Au4f lines, the detection limits would be much worse for the same acquisition time. Detection limits are often quoted as 0.1–1.0% atomic percent for practical analyses, but lower limits can be achieved in many circumstances. Degradation depends on the sensitivity of the material to the wavelength of the X-rays used, the total dose of X-rays, the surface temperature, and the level of vacuum. Metals, alloys, ceramics, and most glasses are not measurably degraded by either non-monochromatic or monochromatic X-rays. However, some, but not all, polymers, catalysts, certain highly oxygenated compounds, various inorganic compounds, and fine organic materials will be affected. Non-monochromatic X-ray sources produce a significant amount of high-energy braking radiation (1–15 keV energy) that directly degrades the surface chemistry of various materials (Turner & Jobory, 1962). Non-monochromatic X-ray sources also generate a considerable amount of heat (100–200 °C) on the surface of the sample, since the anode that generates the X-rays is typically only 1–5 cm from the sample. In conjunction with bremsstrahlung, this heat increases the extent and rate of decomposition of certain materials. Monochromatized X-ray sources located farther (50–100 cm) from the sample do not produce noticeable heat effects. In these sources, a quartz monochromator system diffracts the bremsstrahlung from the X-ray beam, which means that the sample is exposed to only a narrow band of X-ray energy. For example, if aluminum K-alpha X-rays are used, the intrinsic energy band has an Full Width at Half Maximum (FWHM) of 0.43 eV, centered at 1,486.7 eV ($E/\Delta E = 3{,}457$). Using magnesium K-alpha X-rays, the intrinsic energy band has an FWHM of 0.36 eV, centered on 1,253.7 eV ($E/\Delta E = 3{,}483$). These are the intrinsic X-ray linewidths; the energy range to which the sample is exposed depends on the quality and optimization of the X-ray monochromator. As vacuum removes various gasses (e.g., O_2, CO, etc.) and liquids (e.g., water, alcohol, solvents, etc.) originally trapped in or on the surface of the sample, the chemistry and morphology of the surface continue to change until the surface reaches a steady state. This type of decomposition is sometimes difficult to detect (Photoelectron Spectroscopy, 2005) The area measured depends on the design of the instrument. The minimum range of analysis is between 10 and 200 μm. The largest size for a monochromatic X-ray beam is 1–5 mm. Non-monochromatic beams are 10–50 mm in diameter. With the latest XPS imaging instruments using synchrotron radiation as the X-ray source, spectroscopic image resolution of 200 nm or less has been achieved. The instruments are suitable for small (mm range) and large samples (cm range), such as wafers. The limiting factor is the sample holder design, sample transport, and vacuum chamber size. Typically 1–20 minutes for a broad overview scan measuring the amount of all detectable elements, typically 1–15 minutes for a high-resolution scan revealing chemical state differences (multiple passes of the area of interest are often required for a high signal-to-noise ratio for the count result), 1–4 hours for a depth profile measuring 4–5 elements as a function of etch depth (this process time can vary the most as many factors are involved) (Gumerova & Rompel, 2018).

XPS only detects electrons that actually escaped from the sample into the vacuum of the instrument. To escape from the sample, a photoelectron must travel through the sample. Emitted photoelectrons can undergo inelastic collisions, recombination, excitation of the sample, recapture, or trapping in various excited states within the material, all of which can reduce the number of escaping photoelectrons. These effects manifest as an exponential attenuation function with increasing depth, resulting in signals detected by analytes at the surface being much stronger than signals detected by analytes deeper below the sample surface. Thus, the signal measured by XPS is an exponentially surface-weighted signal, and this fact can be used to estimate analyte depth in layered materials (Riga & Judovits, 2001). The ability to extract information about the chemical state, i.e., the local bonding environment, of a particular atomic species from the top nanometers of the sample makes XPS a unique and valuable tool for understanding surface chemistry. The local bonding environment is affected by the formal oxidation state, the identity of the nearest neighbor atoms, and the bond hybridization to the nearest or next-next neighbor atoms. For example, while the nominal binding energy of the C1s electron is 284.6 eV, subtle but reproducible shifts in the actual binding energy, called the chemical shift, provide information about the chemical state (Brandon & Kaplan, 2008). Chemical state analysis is often used for carbon. It shows the presence or absence of the chemical states of carbon in the approximate order of increasing binding energy, such as carbide (-C2-), silane (-Si-CH3), methylene/methyl/hydrocarbon (-CH2-CH2-, CH3-CH2-, and –CH=CH-), amine (-CH2-NH2), alcohol (-C- OH), ketone (-C=O), organic ester (-COOR), carbonate (-CO32-), monofluorocarbon (- CFH -CH2-), difluorocarbon (-CF2-CH2-), and trifluorocarbon (-CH2-CF3), to name a few.

Analysis of the chemical state of the surface of a silicon wafer reveals chemical shifts due to various formal oxidation states, such as n-doped silicon and p-doped silicon (metallic silicon), silicon suboxide (Si_2O), silicon monoxide (SiO), Si_2O_3, and silicon dioxide (SiO_2). An example of this can be seen in the figure "High-resolution spectrum of an oxidized silicon wafer in the energy range of the Si 2p signal" (Naumann D'Alnoncourt et al., 2014). The spectrum of an XPS experiment is a plot of emission intensity versus binding energy. This allows the identification of the elements on the surface based on the unique binding energy of each element. The peak areas in these spectra can also be used to determine the concentration of the elements on the surface. Detailed information on the interpretation of XPS spectra is planned for a future module. In short, the spectrum of an XPS experiment is a plot of emission intensity versus binding energy. In this way, the elements on the surface can be identified based on the unique binding energy of each element. The peak areas in these spectra can also be used to determine the concentration of the elements on the surface. Detailed information on the interpretation of XPS spectra is planned for a future module (Surfaces Editorial Office, 2020). Despite the many advantages of XPS, nothing is foolproof, and nothing is without limitations. The smallest analytical range that XPS can measure is ~10 um. Samples for XPS must be compatible with

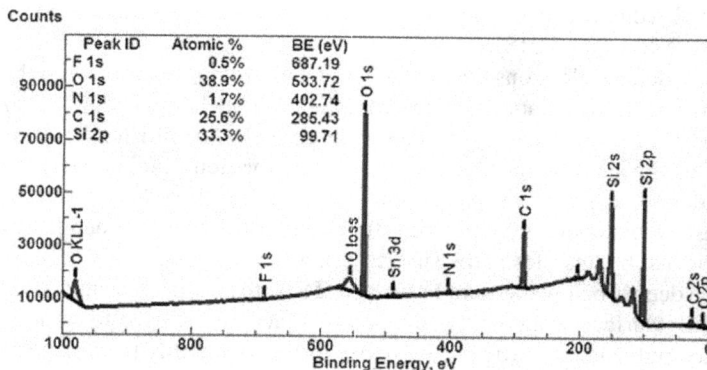

FIGURE 4.5 XPS spectra. Graph showing the binding energies of electrons from different orbitals (F1s, O1s, Si2p, etc.) and their intensities which tell the atomic composition of the sample based on the amounts of each electron from different orbitals present. (https://en.wikipe dia.org/wiki/X-ray_photoelectron_spectroscopy. Figure courtesy of the creative commons license.)

the ultra-high vacuum environment. Since XPS is a surface technique, XPS can only provide a limited amount of organic information. XPS is limited to measurements of elements with an atomic number of 3 or more, so hydrogen or helium cannot be detected. The preparation of XPS spectra is also very time-consuming. The use of a monochromator can also reduce the time per experiment. XPS has a wide range of applications than UPS since it can penetrate the core electrons. XPS can be used to identify all but two elements, determine the chemical state of surfaces and per-form quantitative analyses (Jablonski & Powell, 1999). XPS can detect the diffe-rence in the chemical state between samples. XPS is also able to distinguish between oxidation states of molecules. In addition, XPS can be used to determine chemical shifts. This is because the binding energy does not only depend on the shell of the electron. It also depends on the environment, i.e., the bonds involving the atoms in question. Therefore, a primary carbon has a somewhat different binding energy than a carboxylic carbon, as shown in the example in Figure 4.5.

The peaks from the XPS spectra indicate the relative number of electrons with a given binding energy. The shorter the peak, the fewer electrons are represented. For example, if a peak A is only half as high as another peak B, it means that only half as many electrons with the binding energy of A were detected as with the binding energy of B. Therefore, peak intensities provide information about the percentage compos-ition of a material. As can be seen in the figure above, O1s have the largest peak, which means that the atomic composition of oxygen is the largest. The greater the binding energy, the more strongly the electron is attracted to the nucleus, meaning that the peaks of 1s electrons have higher energy than the peaks of 2s electrons. Electrons in 2s have higher energy than those in 2p. Figure 4.5 shows examples of these standards to help interpret the spectra. Some instruments have functions to identify peaks, but otherwise, the identification of peaks/lines in the spectra can be completed by looking at standards (Hävecker et al., 2012).

4.4 APPLICATION

XPS is very well suited for surfaces. This is because the kinetic energy of the escaping photoelectrons limits the depth that can be studied. The samples studied are all types of solids, from metals to frozen liquids. When the sample is irradiated, the electrons are ejected from the inner shells of the atoms (Turner & Jobory, 1962). There are several areas suitable for measurement by XPS: elemental composition, empirical formula determination, chemical state, electronic state, binding energy, and layer thickness at the top of surfaces. Some specific examples of systems studied by XPS are the analysis of stains and residues on surfaces, reactive frictional wear in solid state reactions, the thickness of silicon oxide, and measurements of dosage (Ellis et al., 2005).

4.4.1 DEPTH PROFILING

Depth profiling uses a sputter source. Successive layers are removed from the surface of a sample so that element depth profiles can be recorded in the near-surface region (Turner & Jobory, 1962). This is useful in the composition of thin film angle-dependent measurements: When the measurement angle is changed, the depth of the collected information can be varied by 1–10 nm. A special imaging mode can be used to determine the distribution of elements in surface structures. This technique is useful for dimensions up to about 3 μm, the elemental composition of the surface (usually uppermost 1–10 nm), the empirical formula of pure materials, elements contaminating a surface, the chemical or electronic state of each element in the surface, the uniformity of elemental composition across the entire surface (or line profiling or mapping) (Surfaces Editorial Office, 2020), and uniformity of elemental composition as a function of ion beam etching (or depth profiling).

4.5 SUMMARY

XPS is one of the best-known and most widely used surface analysis techniques, and applications of this method continue to grow. The two main advantages of this technique are its surface sensitivity (~10 nm) and its ability to detect differences in the chemical environment. These properties have uniquely enabled XPS to answer many research questions. XPS can detect all elements except hydrogen and helium with detection limits of about 0.1–1%, is quantitative without standards, and is relatively nondestructive. The lateral resolution of a few micrometers to hundreds of micrometers can be informative. Since XPS is extremely surface sensitive, care must be taken not to contaminate the surface. Charging of insulating samples can cause problems, but charge neutralizers help solve the problem. XPS remains a very popular tool for surface analysis, but care must be taken when interpreting the data.

REFERENCES

Brandon, D. G., & Kaplan, W. D. (2008). *Microstructural Characterization of Materials*. John Wiley.

de Vera, P., & Garcia-Molina, R. (2019). Electron inelastic mean free paths in condensed matter down to a few electronvolts. *Journal of Physical Chemistry C, 123*(4), 2075–2083. https://doi.org/10.1021/acs.jpcc.8b10832

Ellis, A. M., Feher, M., & Wright, T. G. (2005). *Electronic and Photoelectron Spectroscopy: Fundamentals and Case Studies* (Vol. 9780521817370). Cambridge University Press. https://doi.org/10.1017/CBO9781139165037

Gumerova, N. I., & Rompel, A. (2018). Synthesis, structures and applications of electron-rich polyoxometalates. *Nature Reviews Chemistry, 2*(2). https://doi.org/10.1038/s41570-018-0112

Hävecker, M., Wrabetz, S., Kröhnert, J., Csepei, L. I., Naumann D'Alnoncourt, R., Kolen'Ko, Y. v., Girgsdies, F., Schlögl, R., & Trunschke, A. (2012). Surface chemistry of phase-pure M1 MoVTeNb oxide during operation in selective oxidation of propane to acrylic acid. *Journal of Catalysis, 285*(1), 48–60. https://doi.org/10.1016/j.jcat.2011.09.012

Huang, H., & Zhang, J. (2022). *Surface Functionalization for Heterogeneous Catalysis.* Reference Module in Materials Science and Materials Engineering, https://doi.org/10.1016/B978-0-12-822425-0.00073-7

Jablonski, A., & Powell, C. J. (1999). Relationships between electron inelastic mean free paths, effective attenuation lengths, and mean escape depths. *Journal of Electron Spectroscopy and Related Phenomena, 100*. www.elsevier.nl/locate/elspec

Li, Y., He, Y., Qiu, J., Zhao, J., Ye, Q., Zhu, Y., & Mao, J. (2018). Enhancement of pitting corrosion resistance of austenitic stainless steel through deposition of amorphous/nanocrystalline oxy-nitrided phases by active screen plasma treatment. *Materials Research, 21*(6). https://doi.org/10.1590/1980-5373-mr-2017-0697

Naumann D'Alnoncourt, R., Csepei, L. I., Hävecker, M., Girgsdies, F., Schuster, M. E., Schlögl, R., & Trunschke, A. (2014). The reaction network in propane oxidation over phase-pure MoVTeNb M1 oxide catalysts. *Journal of Catalysis, 311*, 369–385. https://doi.org/10.1016/j.jcat.2013.12.008

Photoelectron Spectroscopy. (2005). www.springer.de/phys/

Rahmayeni, A. A., Stiadi, Y., Lee, H. J., & Zulhadjri . (2019). Green synthesis and characterization of ZnO-CoFe$_2$O$_4$ semiconductor photocatalysts prepared using Rambutan (*nephelium lappaceum* L.) peel extract. *Materials Research, 22*(5). https://doi.org/10.1590/1980-5373-MR-2019-0228

Rawat, P. S., Srivastava, R. C., Dixit, G., & Asokan, K. (2020). Structural, functional and magnetic ordering modifications in graphene oxide and graphite by 100 MeV gold ion irradiation. *Vacuum, 182*. https://doi.org/10.1016/j.vacuum.2020.109700

Ray, S., & Shard, A. G. (2011). Quantitative analysis of adsorbed proteins by X-ray photoelectron spectroscopy. *Analytical Chemistry, 83*(22), 8659–8666. https://doi.org/10.1021/ac202110x

Riga, A. T., & Judovits, L. (2001). *Materials characterization by dynamic and modulated thermal analytical techniques.* ASTM (online).

Siegbahn, K., & Edvarson, K. (1958). X-ray spectroscopy in the precision range of 1: 10 s. *Nuclear Physic, 1*, 452–459.

Surfaces Editorial Office. (2020). Acknowledgement to reviewers of Surfaces in 2019. *Surfaces, 3*(1), 48–49. doi:10.3390/surfaces3010005

Turner, D. W., & Jobory, M. I. A. L. (1962). Determination of ionization potentials by photoelectron energy measurement [1]. *The Journal of Chemical Physics, 37*(12), 3007–3008. https://doi.org/10.1063/1.1733134

Vashishtha, N., & Sapate, S. (2018). Effect of experimental parameters on wear response of thermally sprayed carbide based coatings. *Materials Research*, *22*(1). https://doi.org/10.1590/1980-5373-MR-2018-0475

Voiry, D., Shin, H. S., Loh, K. P., & Chhowalla, M. (2018). Low-dimensional catalysts for hydrogen evolution and CO_2 reduction. *Nature Reviews Chemistry*, *2*(3). https://doi.org/10.1038/s41570-017-0105

5 Scanning Electron Microscope (SEM)

Scanning electron microscope (SEM) is an instrument that uses electron microscopy technique by focusing a beam of electrons on the surface. The electron interaction with the atoms of the sample under investigation generates varying signals that are captured as a function of position and intensity to produce an image (Wikipedia contributors, 2022). SEM is an advancement over the traditional optical microscopy (OM) techniques that offer 400–1,000 times magnification in comparison to 300,000 times magnification offered by a SEM instrument. Further, SEM imaging is dependent on electron interaction with the sample under investigation whereas the OM imaging is totally dependent on light falling on the sample where the image is magnified through a set of lenses. Although SEM images have a higher depth of field when compared to the OM images, these images are grey colored and do not reveal anything about the true color of the sample (Abdullah & Mohammed, 2019). However, the atomic level interaction enhances the elemental details that can be obtained from the SEM images. The SEM technique has wide range of applications in various sectors such as thin films, tribology, and failure analysis of cutting tools to name a few (Grzesik, 2000; Kržan et al., 2009; Navinšek et al., 1995; Nohava et al., 2015; Rech, 2006). The process provides high magnification which in combination with energy dispersive X-ray spectroscopy (EDS) could provide qualitative and semi-quantitative information regarding the material under investigation (Thermo Fisher Scientific Inc. Electron Microscopy Solutions, n.d.). However, there are other attachments and electron detectors such as electron backscatter diffraction (EBSD) setup, X-ray detector, secondary electron detector, and backscattered electron detector that can be helpful for different applications and will be discussed in the subsequent sections. Taking into consideration the vast application of SEM in different fields such as material science, medical science, geology, biology, etc., it is necessary to understand the technique more precisely, and thus the present chapter will discuss the working, construction, and application of the SEM verbosely.

5.1 CONSTRUCTION

Figure 5.1 shows a Zeiss-made EVO 18 SEM with the display unit, EDS attachment and navigation system. However, a generalized SEM consists of several complex lens

DOI: 10.1201/9781003340546-6

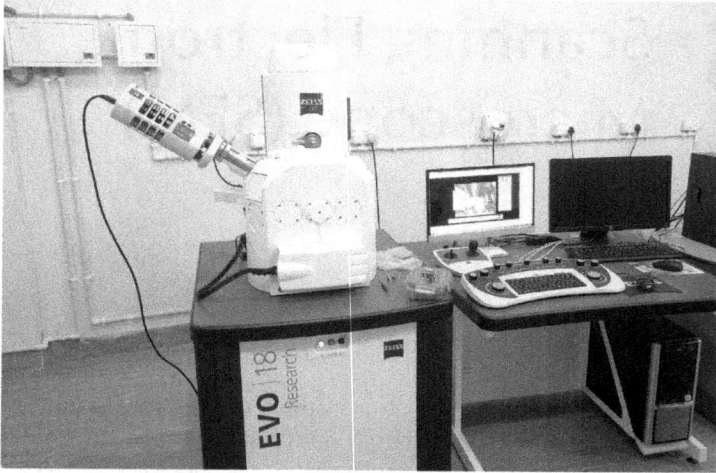

FIGURE 5.1 Zeiss EVO 18 scanning electron microscope. (https://en.wikipedia.org/wiki/File:CETB_Scanning_Electron_Microscope.jpg#filelinks. This work is licensed under the Creative Commons Attribution-ShareAlike 4.0 License.

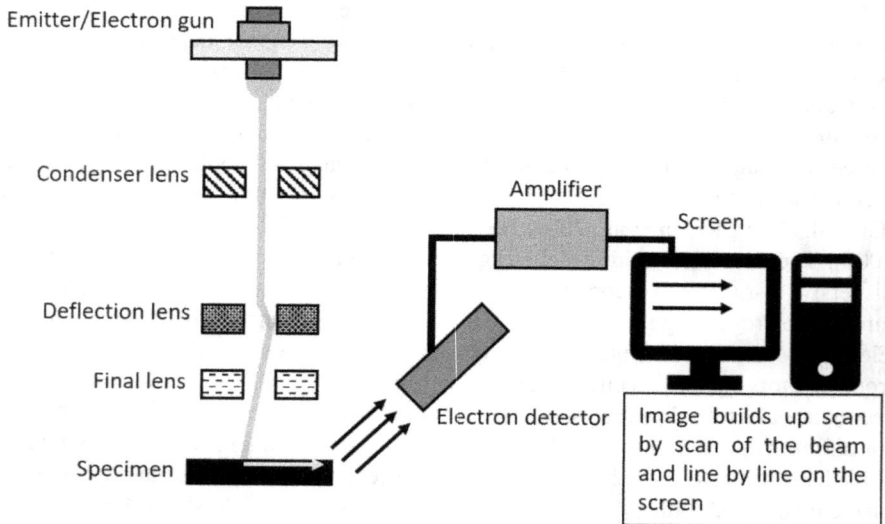

FIGURE 5.2 Schematic representation of SEM components.

arrangement for condensing and focusing the electrons on the sample. Figure 5.2 shows a schematic representation of various components in SEM. A basic SEM setup consists of an electron gun (see Figure 5.3) for generation of high energy electrons. In a SEM setup, the electron gun generates the high energy electrons by thermionic

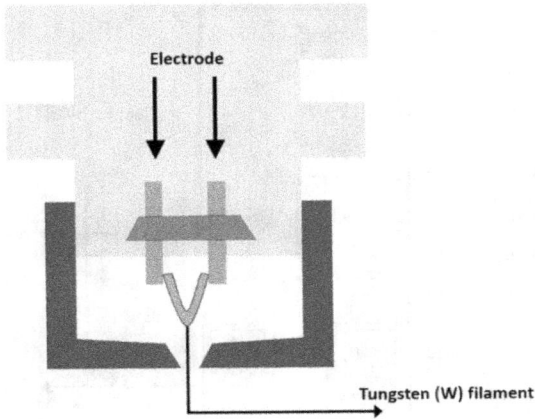

FIGURE 5.3   Schematic representation of electron gun with hairpin filament.

emission using a tungsten filament which is also termed as hairpin filament due to its shape. The electrons emitted from the electron gun would travel through a chamber that consists of a pump to create vacuum. This chamber is facilitated with various lens arrangements to deflect and focus the electron beam on the sample. The sample has a separate chamber for mounting. The SEM system also consists of cameras for visually inspecting the position of sample and the filament within the SEM chamber. To understand the interaction of electrons with the atoms of the sample, different electron detectors are attached to the equipment. Also, the SEM is equipped with a computer display and monitoring system with a high-end software specially designed for SEM (basically developed by the manufacturer) to visualize the camera output and the images during the scanning process. Also, these microscopes mostly have an EDS setup attached for performing elemental analysis (Abdullah & Mohammed, 2019; Hardness & Techniques, 1951).

5.2   WORKING PRINCIPLE

Figure 5.4 shows the schematic representation of a SEM machine. The electron beam having energy in the range of 0.5 to 30 keV is generated from the electron gun which is then focused to a spot of exceedingly small diameter in the range of 5 to 10 nm using condenser lens. Then the condensed beam passes through a set of deflector coils situated in the final lens whose function is to deflect the beam in the x and y axes. This deflection account to scans in a raster fashion over a rectangular area on the sample surface (McMullan, 1995; Wikipedia contributors, 2022). The movement of the scan coils generate the image by point-by-point scanning of the surface of the specimen. Further, higher magnification is achieved by deflecting the electron beam over a smaller area which is further affected by the working distance between the lens and the specimen.

To understand the imaging process using SEM, it is necessary to comprehend the penetration levels of the electrons within a sample or specimen, and the different

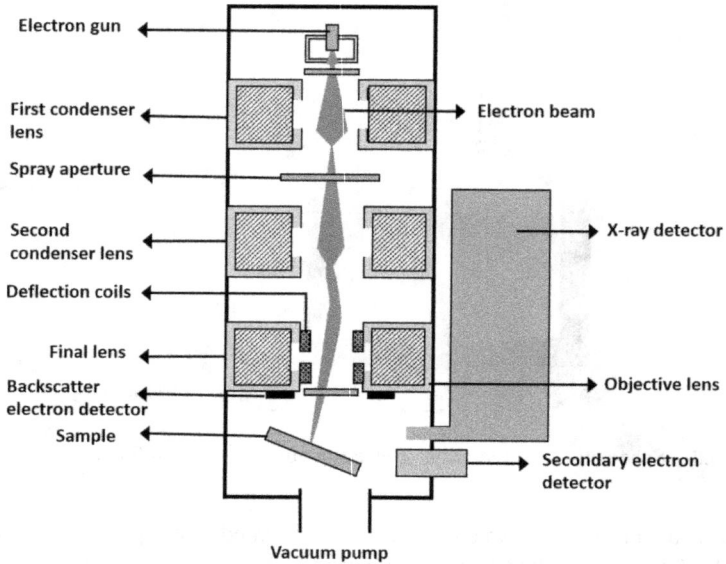

FIGURE 5.4   Schematic representation of scanning electron microscope (SEM).

signals generated due to this interaction. Figure 5.5 shows different signals generated due to electron-specimen interaction. When the electrons come in contact of the surface of the sample, they start losing energy by scattering and absorption which takes place within a teardrop-shaped region called as interaction volume. Further, this interaction leads to generation of various signals like X-rays, backscattered electrons, secondary electrons, Auger electrons and so on. However, only backscattered electrons (BSE) and secondary electrons (SE) are used for image generation in SEM. The direction of positive voltage to the collector screen would result in collection of both BSE and SE signals whereas a negative voltage would account to collection of only BSE signals. The secondary electrons give precise topographical information whereas the backscattered electrons give information about atomic numbers and phase differences. Further, the accelerating voltages significantly impacts the information collected from the sample. Higher accelerating voltages lead to more penetration into the sample whereas lower accelerating voltages gives more information of the surface. Thus, the accelerating voltages must be selected judiciously (Abdullah & Mohammed, 2019; Hardness & Techniques, 1951; Wikipedia contributors, 2022).

5.3   SAMPLE PREPARATION

The samples must be cleaned properly before introducing them into the SEM chamber. Also, there should not be any loose dirt or foreign particles that may stick inside the chamber to cause fault in the machine. Further, the samples should be prepared to fit inside the SEM chamber. However, there are certain mounting techniques that may accommodate samples of different sizes, but the sample preparation must be done

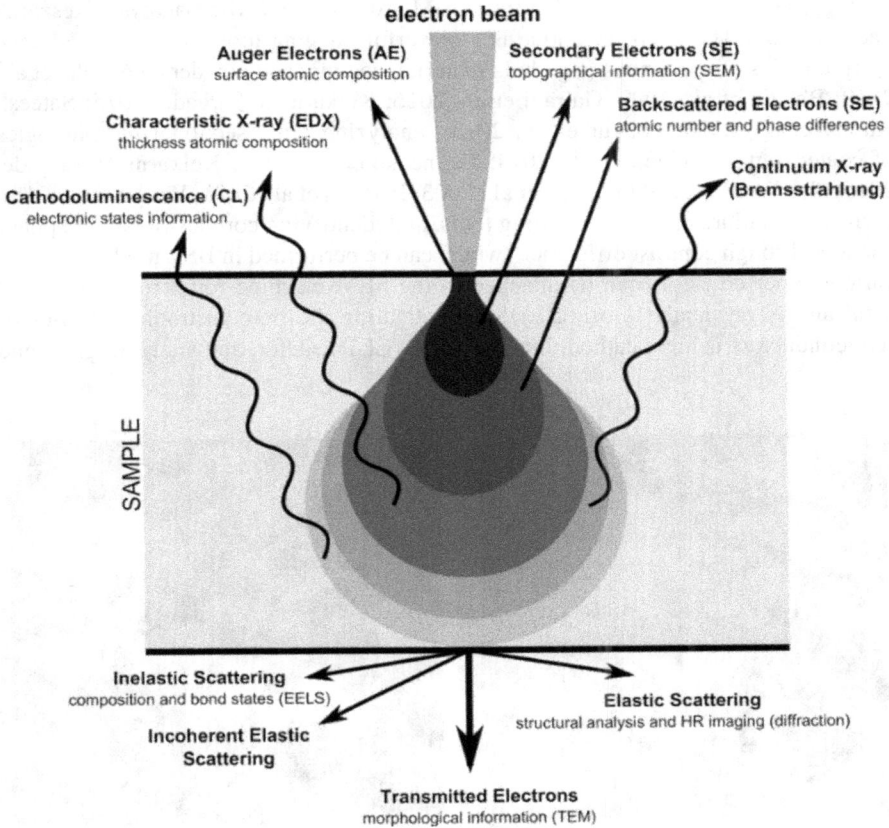

FIGURE 5.5 Interaction volume and types of signals generated due to electron-sample interaction. (https://commons.wikimedia.org/wiki/File:Electron_Interaction_with_Matter.svg. This file is licensed under the Creative Commons Attribution-Share Alike 4.0 International license.)

according to the mounting technique and the maximum sample size specified by the equipment manufacturer. In general, the samples should be electrically conductive so that the electron beam may penetrate through the sample. Purely non-conductive samples would accumulate charge especially in secondary electron mode leading to blurred, non-focused, and faulty images. Thus, it is necessary to have the sample to be conductive at least at the surface for getting better imaging experience with SEM. So, the non-conductive samples are coated with thin layers of gold or platinum to make the surface conductive (Wikipedia contributors, 2022).

5.4 APPLICATION

The SEM technique is one of the most widely used imaging methods in different fields of science and engineering such as material science, forensic science, biology, geology, and medical science (Australian Microscope and Microanalysis Research Facility, n.d.). The SEM technique is a powerful imaging tool for studying surface morphologies of specimens (e.g., bulk material, coating, and powders) (Anidha et al., 2019; Bobzin et al., 2009; Gajrani et al., 2016; Kulkarni & Sargade, 2015; Sateesh Kumar et al., 2020; Thakur et al., 2015), analyzing cross-sectional morphologies of coated samples (Jeon et al., 2014; Keunecke et al., 2010; Kulkarni & Sargade, 2015; Nordin et al., 1999; Panjan et al., 2003; Persson et al., 2001; Yang et al., 2016), performing failure analysis of cutting tools, and identifying compositional and phase changes through contrast difference (which can be performed in BSE mode). Further, various other detectors can be attached to the SEM machine for performing elemental and X-ray analysis using EDS, investigating the optoelectronic behavior of semiconductors using a cathodoluminescence (CL) detector, and studying grain and

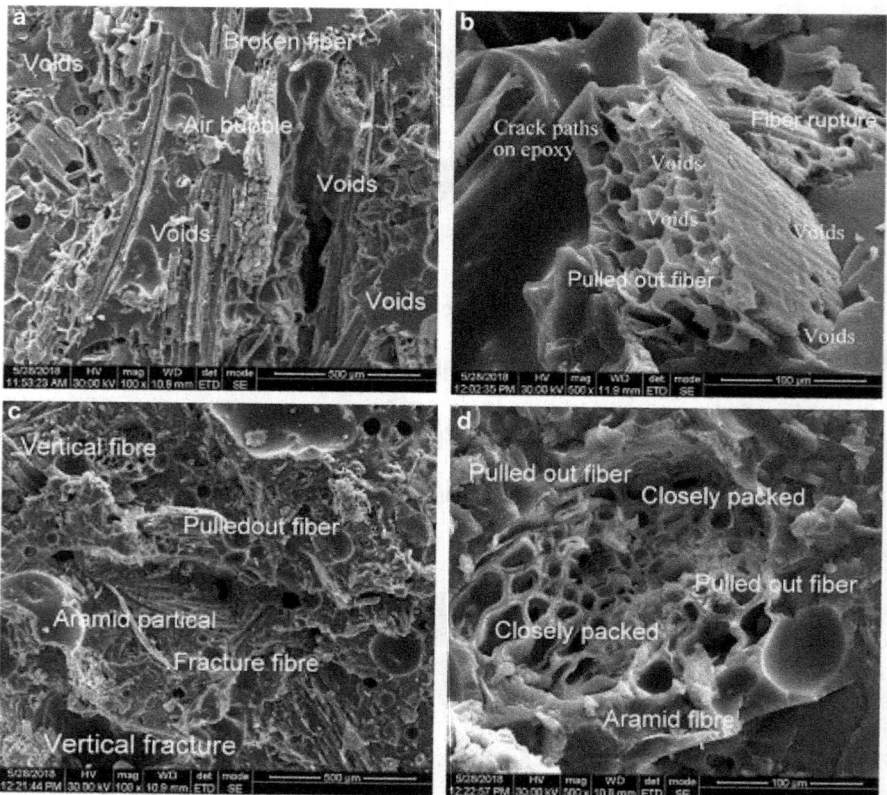

FIGURE 5.6   SEM micrographs showing (a and b) tensile fractured specimen of untreated BG/E composite and (c and d) tensile fractured specimen of treated AF composite. (For details, see Anidha, S., Latha, N., & Muthukkumar, M. (2019). Reinforcement of Aramid fiber with bagasse epoxy bio-degradable composite: Investigations on mechanical properties and surface morphology. *Journal of Materials Research and Technology*, 8(3), 3198–3212. https://doi.org/10.1016/j.jmrt.2019.05.008. Reprinted with permission from Elsevier.)

crystallographic orientation with the help of EBSD detector (Australian Microscope and Microanalysis Research Facility, n.d.).

Figure 5.6 shows the SEM surface morphology of tensile fractured specimen of untreated bagasse/epoxy (BG/E) composite and treated Aramid fiber (AF) composite, respectively (Anidha et al., 2019). The higher depth of field enables the study of various detailed aspects of fractured samples, such as the formation of voids, broken fibers, air bubbles, vertical fractures, pulled-out fibers, and so on. Further, Figure 5.7 shows the study of cross-sectional morphologies of AlTiN-coated cutting tools with different thin-film thicknesses using SEM. The SEM micrographs revealed interface cracks, growth of coating from the substrate to the top surface, and coating thickness (Kumar & Patel, 2018). The SEM in combination with EBSD can also be used to precisely study the grain and crystallographic orientation. Figure 5.8 shows the microstructure of undeformed ferritic stainless steel generated using EBSD. The EBSD

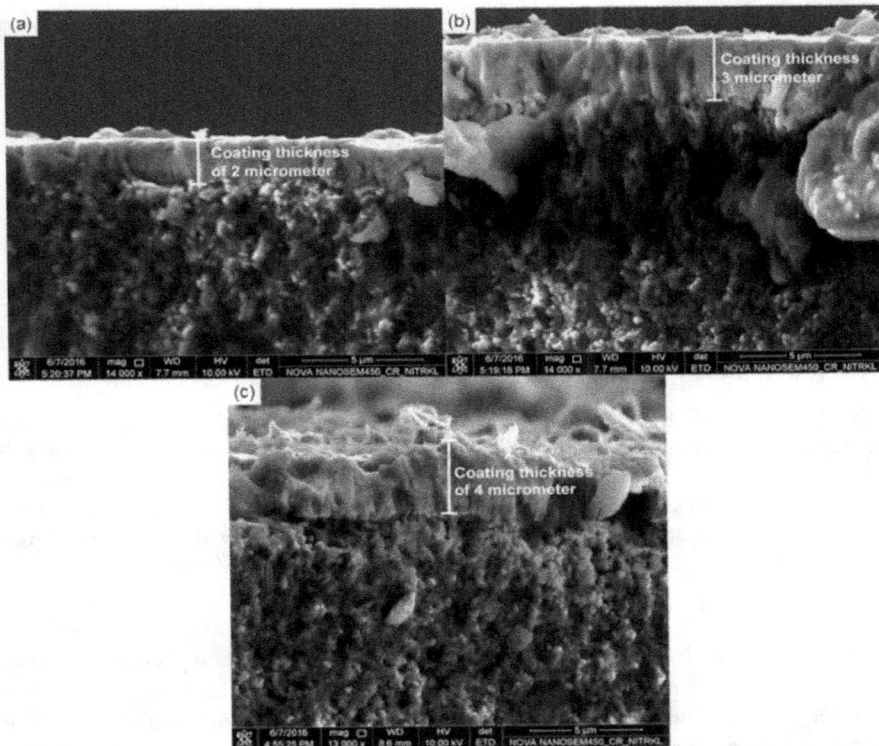

FIGURE 5.7 SEM micrographs showing cross-sectional morphologies of monolayered AlCrN coating at different thin film thicknesses. (For details, see: Kumar, C. S., & Patel, S. K. (2017). Experimental and numerical investigations on the effect of varying AlTiN coating thickness on hard machining performance of Al_2O_3-TiCN mixed ceramic inserts. *Surface & Coatings Technology*, *309*, 266–281. https://doi.org/10.1016/j.surfcoat.2016.11.080. Reprinted with permission from Elsevier.)

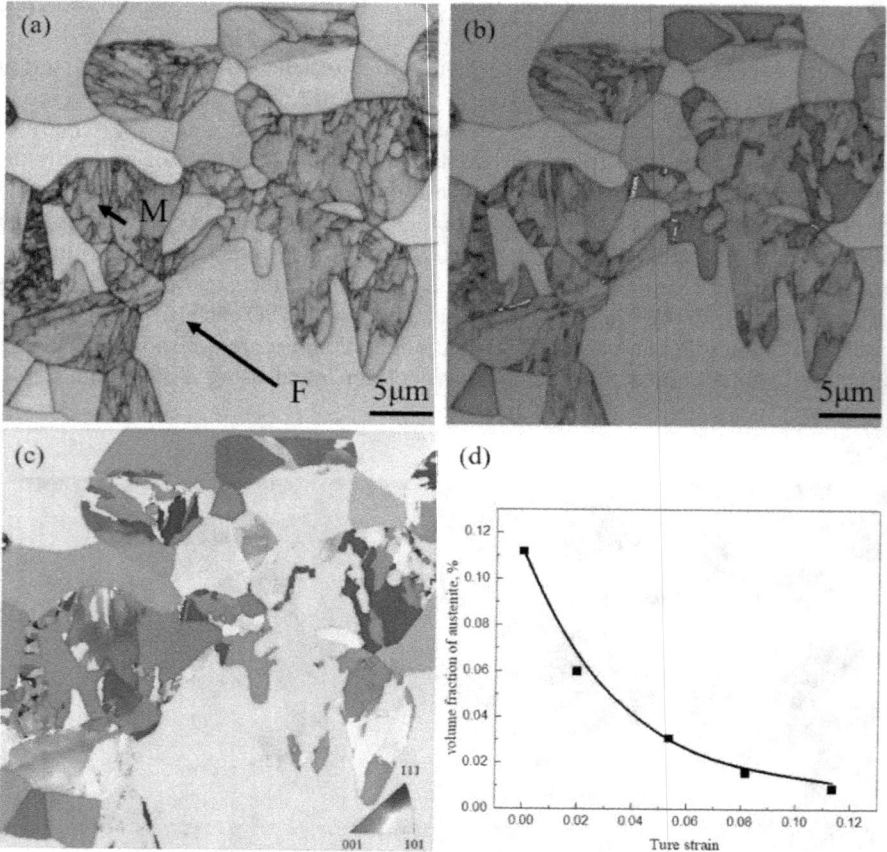

FIGURE 5.8 EBSD analyses of microstructure before deformation (a–c) and relationship between austenite fraction and strain (d). (a) Image quality map, (b) phase map, and (c) IPF (inverse pole figure) map. (For details, see Gao F, Gao Z, Zhu Q, Yu F, & Liu Z, Deformation behavior of retained austenite and its effect on plasticity based on in-situ EBSD analysis for transformable ferritic stainless steel. *Journal of Materials Research and Technology*, https://doi.org/10.1016/j.jmrt.2022.07.160. Reprinted with permission from Elsevier.)

analysis shows the formation of different phases such as ferrite, austenite, and martensite with the lattice defects and other parameters (Gao et al., 2022).

Also, the SEM technique can be a useful tool for conducting failure analysis of materials. One such example is shown in Figure 5.9 where the SEM in combination with EDS was used to study the wear on the AlTiN-coated tools. The EDS helped in carrying out the elemental analysis and was able to give indicative results related to oxidation of cutting tool and coating materials which was validated by X-ray diffraction results (Kumar & Patel, 2018). These are some of the examples that elaborate on the application of SEM.

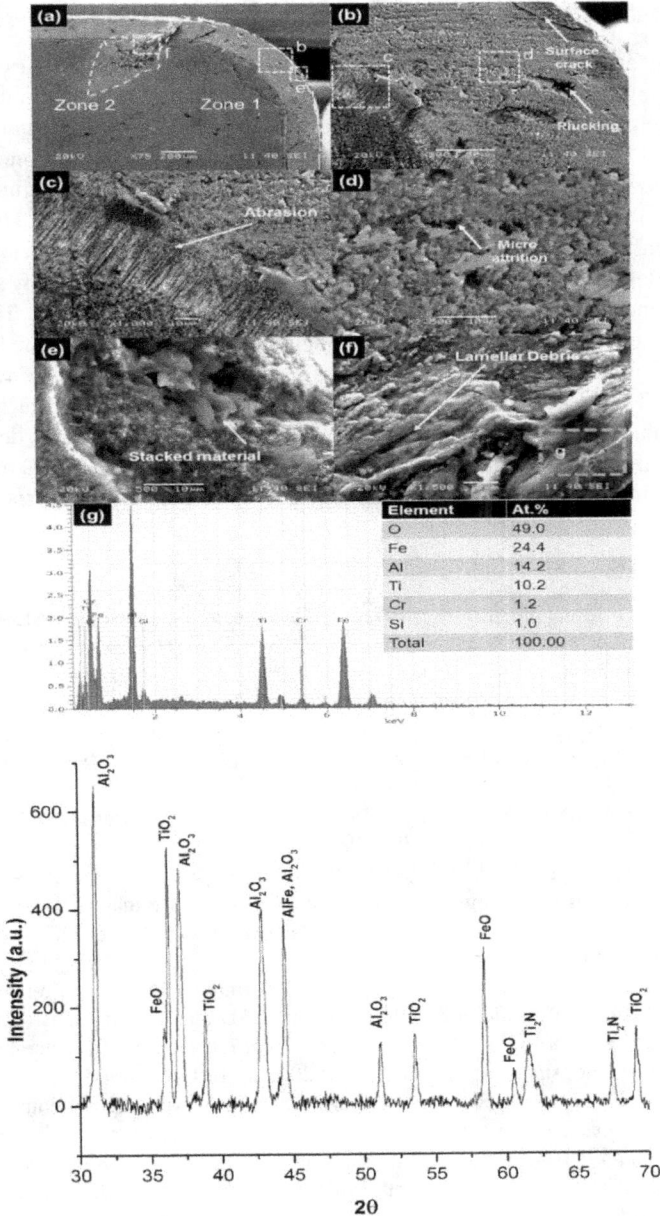

FIGURE 5.9 SEM micrographs and XRD phase analysis of the rake surface of multilayered AlTiN-coated cutting tool after machining. (For details, see Kumar, C. S., & Patel, S. K. (2018). Performance analysis and comparative assessment of nano-composite TiAlSiN/TiSiN/TiAlN coating in hard turning of AISI 52100 steel. *Surface and Coatings Technology*, *335*(September 2017), 265–279. https://doi.org/10.1016/j.surfcoat.2017.12.048 This is an open access article distributed under the terms of the Creative Commons CC-BY license by Elsevier.)

5.5   SUMMARY

SEM is a technique that is used to take high magnification images of various materials that are being used in different fields of research. Some of the common applications of SEM are to study the surface morphology of different materials, analyze wear on cutting tools, determine thin-film thickness, and study crack formation at the micro and nano levels. A generalized SEM setup consists of an electron gun with tungsten filament for generating electrons through thermionic emission. The generated electrons will be in the form of a beam that will pass through a set of lenses to condense and deflect the bean onto the specimen. The image is generated by scanning the scattering and absorption of electrons on the surface of the specimen. The electrons that strike the specimen can penetrate through the sample, leading to the formation of interaction volume. This technique is one of the most widely used imaging techniques incorporated in research for quality control and failure analysis. The technique in combination with different electron and signal detectors such as SE detector, BSE detector, and X-ray detector, and additional attachments such as EBSD and EDS can be used for imaging as well as qualitative and semi-quantitative analysis.

REFERENCES

Abdullah, A., & Mohammed, A. (2019). Scanning electron microscopy (SEM): A review. *Proceedings of 2018 International Conference on Hydraulics and Pneumatics– HERVEX*, 77–85.

Anidha, S., Latha, N., & Muthukkumar, M. (2019). Reinforcement of Aramid fiber with bagasse epoxy bio-degradable composite: Investigations on mechanical properties and surface morphology. *Journal of Materials Research and Technology*, 8(3), 3198–3212. https://doi.org/10.1016/j.jmrt.2019.05.008

Australian Microscope and Microanalysis Research Facility. (n.d.). Retrieved August 8, 2022, from https://myscope.training/#/SEMlevel_3_1

Bobzin, K., Bagcivan, N., Immich, P., Bolz, S., Alami, J., & Cremer, R. (2009). Advantages of nanocomposite coatings deposited by high power pulse magnetron sputtering technology. *Journal of Materials Processing Technology*, 209(1), 165–170. https://doi.org/10.1016/j.jmatprotec.2008.01.035

Gao, F., Gao, Z., Zhu, Q., Yu, F., Liu, Z. (2022). Deformation behavior of retained austenite and its effect on plasticity based on in-situ EBSD analysis for transformable ferritic stainless steel. *Journal of Materials Research and Technology*, 20(September–October), 1976–1992. https://doi.org/10.1016/j.jmrt.2022.07.160

Gajrani, K. K., Reddy, R. P. K., & Sankar, M. R. (2016). Experimental comparative study of conventional, micro-textured and coated micro-textured tools during machining of hardened AISI 1040 alloy steel. *International Journal of Machining and Machinability of Materials*, 18(5–6), 522–539. https://doi.org/10.1504/IJMMM.2016.078982

Grzesik, W. (2000). Influence of thin hard coatings on frictional behaviour in the orthogonal cutting process. *Tribology International*, 33(2), 131–140. https://doi.org/10.1016/S0301-679X(00)00072-4

Hardness, A. I. I., & Techniques, M. (1951). *Appendix A: Techniques for Measuring*. Springer. https://link.springer.com/content/pdf/bbm:978-1-4612-2364-1/1.pdf

Jeon, S., Van Tyne, C. J., & Lee, H. (2014). Degradation of TiAlN coatings by the accelerated life test using pulsed laser ablation. *Ceramics International*, 40(6), 8677–8685. https://doi.org/10.1016/j.ceramint.2014.01.085

Keunecke, M., Stein, C., Bewilogua, K., Koelker, W., Kassel, D., & den Berg, H. van. (2010). Modified TiAlN coatings prepared by d.c. pulsed magnetron sputtering. *Surface and Coatings Technology, 205*(5), 1273–1278. https://doi.org/10.1016/j.surfcoat.2010.09.023

Kržan, B., Novotny-Farkas, F., & Vižintin, J. (2009). Tribological behavior of tungsten-doped DLC coating under oil lubrication. *Tribology International, 42*(2), 229–235. https://doi.org/10.1016/j.triboint.2008.06.011

Kulkarni, A. P., & Sargade, V. G. (2015). Characterization and performance of AlTiN, AlTiCrN, TiN/TiAlN PVD coated carbide tools while turning SS 304. *Materials and Manufacturing Processes, 30*(6), 748–755. https://doi.org/10.1080/10426914.2014.984217

Kumar, C. S., & Patel, S. K. (2018). Surface & coatings technology experimental and numerical investigations on the effect of varying AlTiN coating thickness on hard machining performance of Al 2 O 3-TiCN mixed ceramic inserts. *SCT, 309*, 266–281. https://doi.org/10.1016/j.surfcoat.2016.11.080

McMullan, D. (1995). Scanning electron microscopy 1928–1965. *Scanning, 17*, 175–185.

Navinšek, B., Panjan, P., & Cvelbar, A. (1995). Characterization of low temperature CrN and TiN (PVD) hard coatings. *Surface and Coatings Technology, 74–75*(Part 1), 155–161. https://doi.org/10.1016/0257-8972(95)08214-X

Nohava, J., Dessarzin, P., Karvankova, P., & Morstein, M. (2015). Characterization of tribological behavior and wear mechanisms of novel oxynitride PVD coatings designed for applications at high temperatures. *Tribology International, 81*, 231–239. https://doi.org/10.1016/j.triboint.2014.08.016

Nordin, M., Herranen, M., & Hogmark, S. (1999). Influence of lamellae thickness on the corrosion behaviour of multilayered PVD TiN/CrN coatings. *Thin Solid Films, 348*(1), 202–209. https://doi.org/10.1016/S0040-6090(99)00192-3

Panjan, P., Čekada, M., & Navinšek, B. (2003). A new experimental method for studying the cracking behaviour of PVD multilayer coatings. *Surface and Coatings Technology, 174–175*, 55–62. https://doi.org/10.1016/S0257-8972(03)00618-2

Persson, A., Bergström, J., Burman, C., & Hogmark, S. (2001). Influence of deposition temperature and time during PVD coating of CrN on corrosive wear in liquid aluminium. *Surface and Coatings Technology, 146–147*, 42–47. https://doi.org/10.1016/S0257-8972(01)01366-4

Rech, J. (2006). Influence of cutting tool coatings on the tribological phenomena at the tool-chip interface in orthogonal dry turning. *Surface and Coatings Technology, 200*(16–17), 5132–5139. https://doi.org/10.1016/j.surfcoat.2005.05.032

Sateesh Kumar, C., Majumder, H., Khan, A., & Patel, S. K. (2020). Applicability of DLC and WC/C low friction coatings on Al2O3/TiCN mixed ceramic cutting tools for dry machining of hardened 52100 steel. *Ceramics International, 46*(8), 11889–11897. https://doi.org/10.1016/j.ceramint.2020.01.225

Thakur, A., Gangopadhyay, S., & Mohanty, A. (2015). Investigation on some machinability aspects of Inconel 825 during dry turning. *Materials and Manufacturing Processes, 30*(8), 1026–1034. https://doi.org/10.1080/10426914.2014.984216

Thermo Fisher Scientific Inc. Electron Microscopy Solutions. (n.d.). Retrieved August 8, 2022, from www.thermofisher.com/es/es/home/electron-microscopy/products/scanning-electron-microscopes/apreo-sem.html

Wikipedia contributors. (2022). *Scanning electron microscope – Wikipedia, The Free Encyclopedia*. https://en.wikipedia.org/w/index.php?title=Scanning_electron_microscope&oldid=1100146463

Yang, W., Xiong, J., Guo, Z., Du, H., Yang, T., Tang, J., & Wen, B. (2016). Structure and properties of PVD TiAlN and TiAlN/CrAlN coated Ti(C, N)-based cermets. *Ceramics International, 43*(2), 1911–1915. https://doi.org/10.1016/j.ceramint.2016.10.151

6 Field Emission Scanning Electron Microscope (FESEM)

Field emission scanning electron microscope (FESEM) is an advancement over conventional scanning electron microscope (SEM) that serves higher resolution and imaging capabilities with wide energy ranges. The functioning of FESEM is like a conventional SEM with a single difference, i.e., the technique of generation of the electron beam which will be discussed in the subsequent sections of the chapter. The capability of working at lower accelerating voltages reduces the charging effect on the nonconductive samples. When compared to a conventional SEM that uses a hairpin-shaped tungsten filament wire which is heated to generate electrons, the FESEM uses a tungsten crystal with a sharp pointed end which is called a field emission gun (FEM) that enables the generation of highly focused electrons even with low accelerating voltages. Also, the smaller spot size enables magnification which can be six times more than that of a conventional SEM. Thus, more information can be attained using a FESEM with virtually unlimited depth of field (Australian Microscope and Microanalysis Research Facility, n.d.; Photometrics, n.d.; Universitat Politècnica de València, n.d.). The high-resolution images that can be obtained at low accelerating voltages make FESEM an excellent choice for overly sensitive and nonconductive samples.

FESEM can be used in the field of material science, geology, biology, failure analysis, etc., like a conventional SEM but can provide more information in terms of depth of field, resolution, and magnification. Some common areas of FESEM application are tool wear analysis (Halim et al., 2019; Sateesh Kumar & Kumar Patel, 2017), high-resolution imaging of biological surfaces (Kirk, 2017; Pawley, 1997), the surface morphology of materials (Das et al., 2016) and cross-sectional morphologies of thin-film samples (Chang & Chao, 2021). However, it is necessary to understand the difference between FESEM and conventional SEM so that their choice of application can be decided judiciously. Thus, the present chapter deals with the construction, working, sample preparation, and applications associated with a FESEM.

6.1 CONSTRUCTION

Figure 6.1 illustrates a SIGMA VP FESEM from Zeiss. The components involved in FESEM are similar to that of a conventional SEM. Figure 6.2 shows a schematic

DOI: 10.1201/9781003340546-7

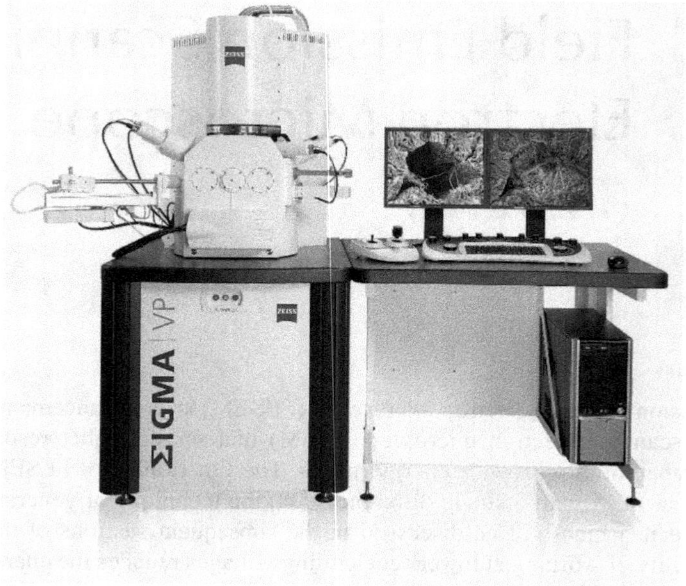

FIGURE 6.1   Zeiss SIGMA VP FESEM. (www.flickr.com/photos/zeissmicro/10710025785/. This work is licensed under the Creative Commons Attribution-CC BY-SA.)

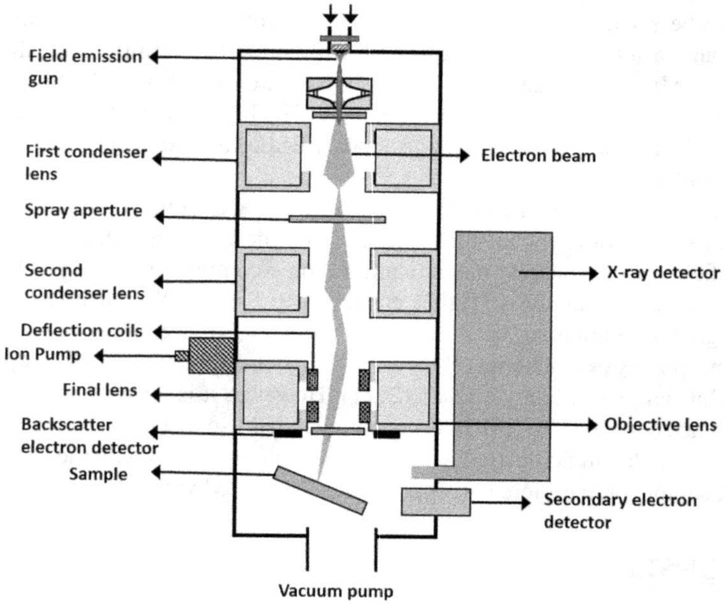

FIGURE 6.2   Schematic representation of field emission scanning electron microscope (FESEM) with electron detectors and lens arrangement.

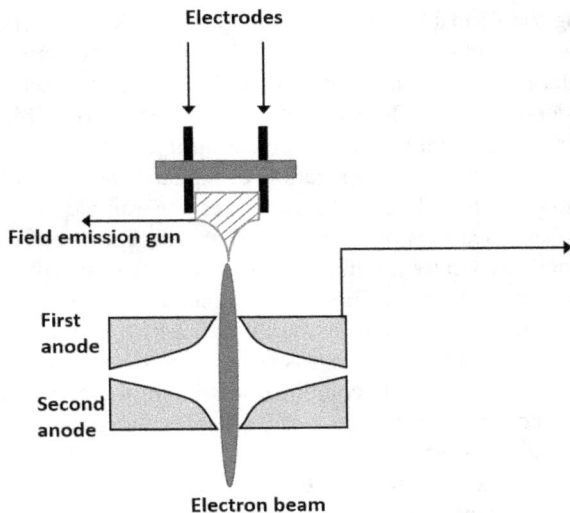

FIGURE 6.3 Schematic representation of field emission gun (FEG) showing sharp tipped tungsten crystal and set of anodes for extracting, accelerating, or deaccelerating the electrons.

representation of FESEM. The setup consists of a chamber in which condenser, deflection, and other lens arrangements are provided to deflect and focus the electron beam onto the sample. The camera and display system are also similar to that of a conventional SEM. However, the electron generation system is changed by replacing the hairpin-shaped tungsten filament wire in the electron gun of conventional SEM with a sharp-tipped tungsten crystal. Figure 6.3 shows a schematic representation of a FEG. The FEG consists of a sharp tungsten crystal with two sets of anodes in front. The first set of anodes is for extracting the electrons whereas the second set of anodes is for accelerating or deaccelerating the electrons that are extracted from the electric field. As this FEG can operate at room temperature, it is often termed as a cold field emission gun (CFEG). The diameter of the tip of the tungsten crystal for CFEG may vary from 10 to 50 nm. This small tip diameter leads to the generation of highly focused electrons with high electric field strength making working possible at low accelerating voltages. The smaller spot size of the focused beam necessitates the use of strict vacuum conditions with vacuum pressures ranging from 10^{-8} to 10^{-9} Pa which can be achieved by using ion getter pumps shown in Figure 6.2. The use of FEG helps in working at low energy levels of 0.02 to 5 keV. However, FESEM is also capable of working at conventional energy ranges from 5 to 30 keV (Photometrics, n.d.; McMullan, 1995; Universitat Politècnica de València, n.d.). Another important feature of FESEM is the use of in-lens detectors which will be discussed in the next section with the working principle.

6.2 WORKING PRINCIPLE

The image formation process for FESEM is the same as that of a conventional SEM where an electron beam strikes the sample leading to scattering and absorption of electrons causing the formation of interaction volume. These generated signals due to the electron–sample interaction are detected by various detectors. For a detailed explanation of the image formation process in SEM, refer to Sections 5.1 and 5.2. However, to understand the difference between FESEM and SEM, it is necessary to again consider the different signals generated during electron–sample interaction. Figure 6.4 shows the schematic illustration of signal generation due to electron–sample interaction. As the electron beam strikes the sample and enters the specimen, the secondary electrons of type 1 (SE1) and backscattered electrons of type 1 (BSE1) are generated locally, whereas the secondary electrons of type 2 (SE2) and backscattered electrons of type 2 (BSE2) are generated around the beam entrance point. Thus, the high-resolution information is carried by SE1 and BSE1 electrons. However, a probe diameter below 1 nm is required to detect these electrons which can be achieved by using a FEG in combination with high-precision electron lenses. Further, for the detection of these type 1 electrons, highly sensitive in-lens detectors are required (McMullan, 1995).

Figure 6.4 shows schematic representation of secondary electrons, backscattered electrons, Auger electrons: AE, cathodoluminescence CL, and X-rays generated during electron–sample interaction. tSE and tBSE indicate the escape depth for SE and BSE, respectively. R is the electron range

An electron detector that can detect secondary electrons of type 2 (SE2) can help in generating an image with a large depth of field, but it would be of either low or

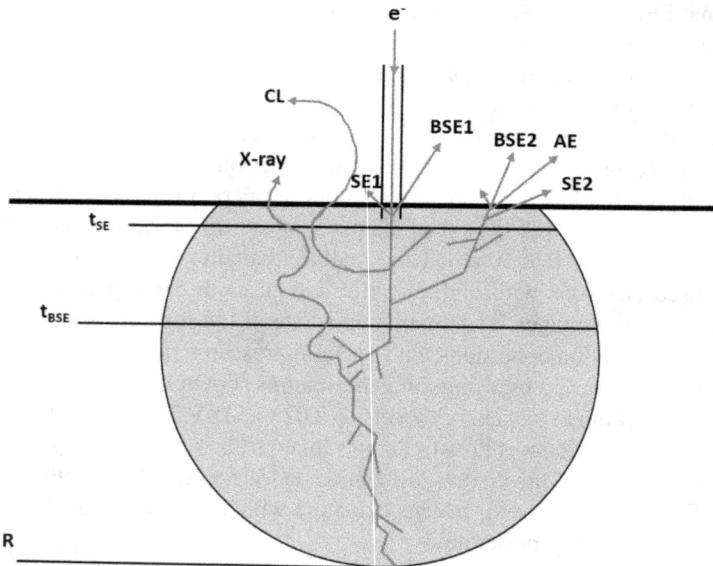

FIGURE 6.4 Schematic representation of secondary electrons, backscattered electrons, Auger electrons: AE, cathodoluminescence CL, and X-rays generated during electron–sample interaction. t_{SE} and t_{BSE} indicate the escape depth for SE and BSE, respectively. R is the electron range.

medium resolution with obvious high accelerating voltages. These images are well suited for performing investigations of samples having large topographic information. However, an in-lens secondary electron detector that is placed inside the electron column can detect low-energy secondary electrons of type 1 (SE1) and thus would provide images that have high resolution. These in-lens detectors can perform very well at low accelerating potentials of below 5 kV. Also, the high sensitivity of these in-lens detectors makes them extremely beneficial for advanced surface characterization studies. Also, low accelerating voltages would prevent the charging effect on nonconductive samples. Further, an out-lens backscattered electron detector that can detect backscattered electrons of type 2 (BSE2) can sense the variation in atomic number and, thus, is widely used to detect the chemical composition changes within a sample. The detector helps in obtaining compositional differences between elements or phases through structural contrast. However, these detected signals can have secondary electron contaminations. This problem of contamination can be solved by using an in-lens backscattered electron detector that can operate at low accelerating voltages leading to the detection of type 1 backscattered electrons (BSE1). The high Z-contrast offered by the detector facilitates the selection of electrons based on their energy. This would help in the identification of elements that are distinguished by a few atoms. Further, these in-lens backscattered electron detectors operate at low voltages which makes them an excellent choice for sensitive and nonconductive samples (Abdullah & Mohammed, 2019; Kirk, 2017; McMullan, 1995; Wikipedia contributors, 2022). Further, energy dispersive spectroscopy (EDS) and electron backscatter diffraction (EBSD) techniques could be implemented for qualitative and quantitative elemental analysis with FESEM in a similar way as was the case with conventional SEM. However, higher magnification and resolution of FESEM in combination with in-lens electron detectors would help in extracting more details from the sample under consideration.

6.3 SAMPLE PREPARATION

The generalized sample preparation for SEM imaging has already been discussed in Section 5.3. As a rule, the samples should be cleaned to avoid any contamination. Further, completely nonconductive samples may still require the application of conductive coating for getting focused images. However, at energies below 5 keV, the charging capacity of the nonconductive materials reduces. Due care must be taken with samples containing moisture. FESEM is a high vacuum process but can be carried out at lower vacuum pressures for moist samples. Alternatively, the samples can be dried or frozen. Further, for performing investigations like studying the grain size, and microstructure or performing EBSD analysis the samples should be properly mounted inside a suitable resin and polished using abrasive, colloidal, or electropolishing techniques based on the requirement. When preparing powder samples, one solution is to stick them properly to the mounting stage using carbon tape. The sample should also be cut so that it may be accommodated inside the sample holder of the FESEM. However, different instrument manufacturers may have different size specifications for the samples (Thermo Fisher Scientific Inc. Electron Microscopy Solutions, n.d.-a, n.d.-b; Wikipedia contributors, 2022).

6.4 APPLICATION

As discussed earlier, FESEM can be used in all the areas of science and engineering where SEM can be used, such as material science, geology, biology, and medical science (Australian Microscope and Microanalysis Research Facility, n.d.). FESEM is one of the most powerful tools for high-resolution imaging of surface morphologies (Halim et al., 2019; Kondo et al., 2020; Koyilada et al., 2016; Thakur & Gangopadhyay, 2016; Xie et al., 2022), cross-sectional morphologies (Koyilada et al., 2016; Wu et al., 2021), biological specimens (Kirk, 2017; Pawley, 1997), failure analysis of cutting tools (Halim et al., 2019; Kumar & Patel, 2018; Zaharah et al., 2014), and so on. In this section, a few examples of the use of FESEM will be discussed. Figure 6.5 shows one of the examples where FESEM has been used to see the high-resolution surface morphology of graphite nanoflakes with different dimensional parameters (Patil et al., 2021). The measured dimensions of the flakes are mentioned in nanometers which is very difficult to measure using a conventional SEM.

Also, FESEM can be used to investigate the failure under different conditions like the wear of cutting tools, and analysis of wear track generated from tribology testing. Figure 6.6 shows the FESEM micrographs showing the wear mechanism on the CrN-coated sample after the pin on disc wear test. The results show abrasion, erosion, groove formation, and metal transfer as some of the wear types on the surfaces under investigation. Further, a close examination of the images also reveals micro-cracks, material pull-out, and large sliding marks in the direction of the movement of the pin (Singh et al., 2022).

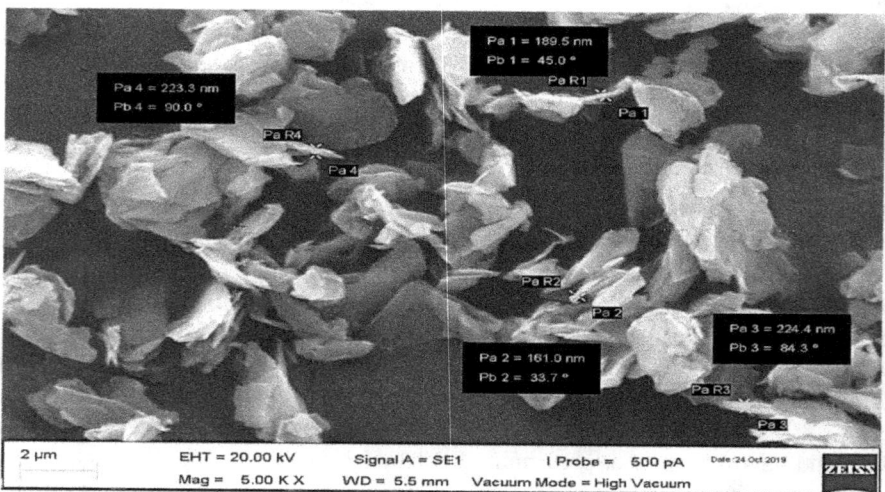

FIGURE 6.5 FESEM micrographs showing graphite nanoflakes. (For details, see Patil, N. A., Pedapati, S. R., Mamat, O., & Lubis, A. M. H. S. (2021). Morphological characterization, statistical modeling and wear behavior of AA7075-titanium carbide-graphite surface composites via friction stir processing. *Journal of Materials Research and Technology, 11,* 2160–2180. https://doi.org/10.1016/j.jmrt.2021.02.054. Reprinted with permission from Elsevier.)

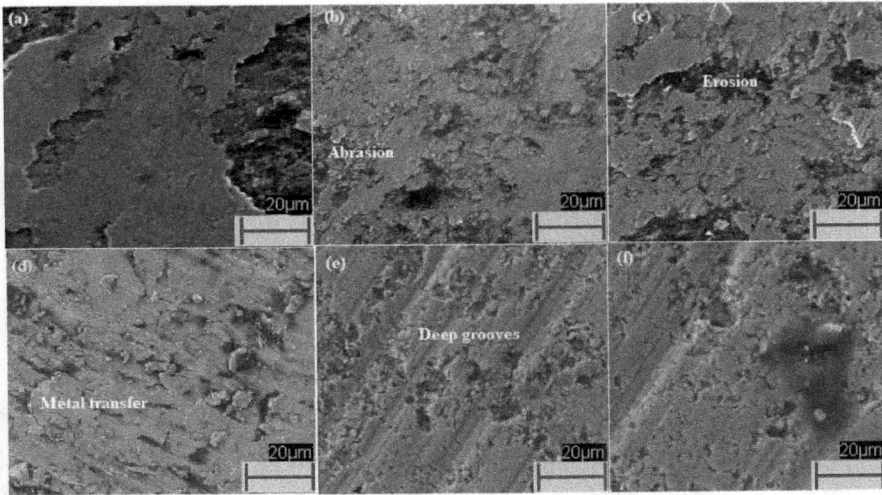

FIGURE 6.6 FESEM micrographs showing wear mechanism on the CrN coated samples along the pin-on-disc wear track. (For details, see Singh, S. K., Chattopadhyaya, S., Pramanik, A., Kumar, S., Pandey, S. M., Walia, R. S., Sharma, S., Khan, A. M., Dwivedi, S. P., Singh, S., & Wojciechowski, S. (2022). Effect of alumina oxide nano-powder on the wear behaviour of CrN coating against cylinder liner using response surface methodology: processing and characterizations. *Journal of Materials Research and Technology, 16,* 1102–1113. https://doi. org/10.1016/j.jmrt.2021.12.062. This is an open access article distributed under the terms of the Creative Commons CC-BY license by Elsevier.)

FESEM is also used extensively for studying the cross-sectional morphologies of coated samples. One such example can be seen in Figure 6.7 where FESEM cross-sectional morphologies of multilayered coatings like AlTiN, AlTiCrSiN–20 V, AlTiCrSiN–100 V, and AlTiCrSiN–180 V coatings have been shown (Chang & Chao, 2021). The figure shows the difference in detailing of FESEM micrographs when the images are taken in backscattering electron and secondary electron mode. On the left, the images are taken using secondary electron imaging that emphasizes more on coating morphology whereas the right side shows the images taken under backscatter electron imaging where compositional changes can be seen throughout the coating structure. It has already been discussed that the detection of backscattered electrons can be used to identify elemental composition that can be visualized through contrast changes.

FESEM can be used for generating high-resolution images of biological specimens that can rival images produced by conventional transmission electron microscopy (TEM) in terms of resolution and contrast (Pawley, 1997). The modern FESEMs have capabilities to study grain boundaries, elemental compositions, and surface morphologies with precise details. Also, the incorporation of in-lens detectors and attachments like EDS and EBSD increases both the qualitative and quantitative capabilities of FESEM.

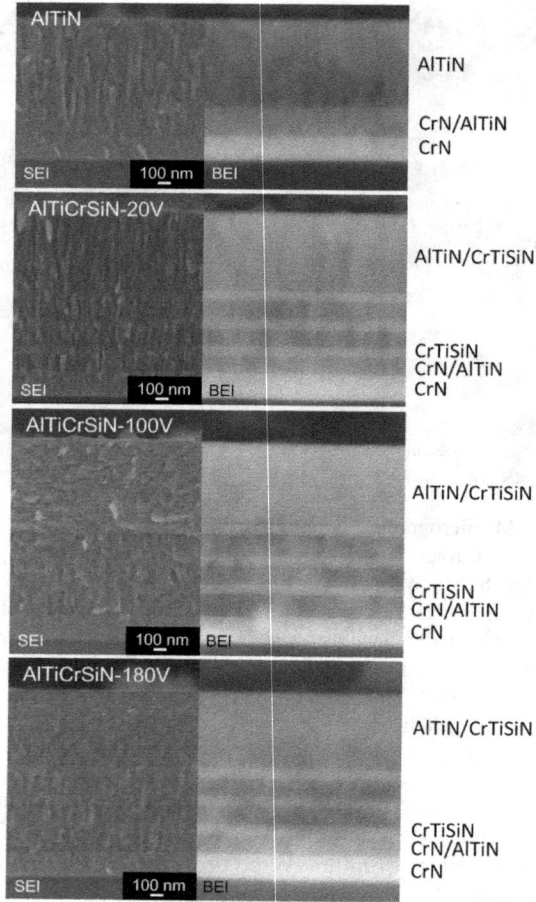

FIGURE 6.7 FESEM cross-sectional images of the AlTiN, AlTiCrSiN–20 V, AlTiCrSiN–100 V, and AlTiCrSiN–180 V coatings. (For details, see Chang, Y. Y., & Chao, L. C. (2021). Effect of substrate bias voltage on the mechanical properties of AlTiN/CrTiSiN multilayer hard coatings. *Vacuum,* 190(January), 110241. https://doi.org/10.1016/j.vacuum.2021.110241. Reprinted with permission from Elsevier.)

6.5 SUMMARY

FESEM is one of the most widely used high-resolution microscopes that finds its application in fields like materials science, geology, biology, etc. FESEM process implements a FEG which increases the electric strength due to the small spot size enabling the use of low accelerating voltages and consequently providing higher magnification and image resolution. Also, the narrow beam diameter facilitates the detection of type 1 secondary electrons and backscattered electrons that contain high-resolution information. Additionally, the FESEM technique is an excellent choice for imaging nonconductive and overly sensitive material samples. However, the FESEM process requires very high vacuum levels. Further, EDS and EBSD techniques could be implemented for qualitative and quantitative elemental analysis with FESEM in a similar way as was the case with conventional SEM. However, higher magnification and resolution of FESEM in combination with in-lens electron detectors would help in extracting more details from the sample under consideration.

REFERENCES

Abdullah, A., & Mohammed, A. (2019). Scanning electron microscopy (SEM): A review. *Proceedings of 2018 International Conference on Hydraulics and Pneumatics–HERVEX*-2018 Conference Proceedings, 77–85.

Australian Microscope and Microanalysis Research Facility. (n.d.). Retrieved August 8, 2022, from https://myscope.training/#/SEMlevel_3_1

Chang, Y. Y., & Chao, L. C. (2021). Effect of substrate bias voltage on the mechanical properties of AlTiN/CrTiSiN multilayer hard coatings. *Vacuum*, *190*(January), 110241. https://doi.org/10.1016/j.vacuum.2021.110241

Das, P., Anwar, S., Bajpai, S., & Anwar, S. (2016). Structural and mechanical evolution of TiAlSiN nanocomposite coating under influence of Si3N4 power. *Surface and Coatings Technology*, *307*, 676–682. https://doi.org/10.1016/j.surfcoat.2016.09.065

Halim, N. H. A., Haron, C. H. C., Ghani, J. A., & Azhar, M. F. (2019). Tool wear and chip morphology in high-speed milling of hardened Inconel 718 under dry and cryogenic CO_2 conditions. *Wear*, *426–427*(January), 1683–1690. https://doi.org/10.1016/j.wear.2019.01.095

Kirk, T. L. (2017). A review of scanning electron microscopy in near field emission mode. *Advances in Imaging and Electron Physics*, *204*. https://doi.org/10.1016/bs.aiep.2017.09.002

Kondo, T., Inoue, H., & Minoshima, K. (2020). Thickness dependency of creep crack propagation mechanisms in submicrometer-thick gold films investigated using in situ FESEM and EBSD analysis. *Materials Science and Engineering A*, *790*, 139621. https://doi.org/10.1016/j.msea.2020.139621

Koyilada, B., Gangopadhyay, S., & Thakur, A. (2016). Comparative evaluation of machinability characteristics of Nimonic C-263 using CVD and PVD coated tools. *Measurement*, *85*, 152–163. https://doi.org/10.1016/j.measurement.2016.02.023

Kumar, C. S., & Patel, S. K. (2018). Performance analysis and comparative assessment of nano-composite TiAlSiN/TiSiN/TiAlN coating in hard turning of AISI 52100 steel. *Surface and Coatings Technology*, *335*(September 2017), 265–279. https://doi.org/10.1016/j.surfcoat.2017.12.048

McMullan, D. (1995). Scanning electron microscopy 1928–1965. *Scanning*, *17*, 175–185.

Patil, N. A., Pedapati, S. R., Mamat, O., & Lubis, A. M. H. S. (2021). Morphological characterization, statistical modeling and wear behavior of AA7075-titanium carbide-graphite surface composites via Friction stir processing. *Journal of Materials Research and Technology*, *11*, 2160–2180. https://doi.org/10.1016/j.jmrt.2021.02.054

Pawley, J. (1997). The development of field-emission scanning electron microscopy for imaging biological surfaces. *Scanning*, *19*(5), 324–336.

Photometrics, Inc. (n.d.). Retrieved August 13, 2022, from https://photometrics.net/field-emission-scanning-electron-microscopy-fesem/

Sateesh Kumar, C., & Kumar Patel, S. (2017). Hard machining performance of PVD AlCrN coated Al2O3/TiCN ceramic inserts as a function of thin film thickness. *Ceramics International*, *43*(16), 13314–13329. https://doi.org/10.1016/j.ceramint.2017.07.030

Singh, S. K., Chattopadhyaya, S., Pramanik, A., Kumar, S., Pandey, S. M., Walia, R. S., Sharma, S., Khan, A. M., Dwivedi, S. P., Singh, S., & Wojciechowski, S. (2022). Effect of alumina oxide nano-powder on the wear behaviour of CrN coating against cylinder liner using response surface methodology: processing and characterizations. *Journal of Materials Research and Technology*, *16*, 1102–1113. https://doi.org/10.1016/j.jmrt.2021.12.062

Thakur, A., & Gangopadhyay, S. (2016). Dry machining of nickel-based super alloy as a sustainable alternative using TiN/TiAlN coated tool. *Journal of Cleaner Production*, *129*, 256–268. https://doi.org/10.1016/j.jclepro.2016.04.074

Thermo Fisher Scientific Inc. Electron Microscopy Solutions. (n.d.-a). Retrieved August 8, 2022, from www.thermofisher.com/es/es/home/electron-microscopy/products/scanning-electron-microscopes/apreo-sem.html

Thermo Fisher Scientific Inc. Electron Microscopy Solutions. (n.d.-b). Retrieved August 13, 2022, from www.thermofisher.com/es/es/home/materials-science/learning-center/applications/sample-preparation-techniques-sem.html

Universitat Politècnica de València. (n.d.). Retrieved August 13, 2022, from www.upv.es/entidades/SME/info/859071normali.html.

Wikipedia contributors. (2022). *Scanning electron microscope – Wikipedia, The Free Encyclopedia*. https://en.wikipedia.org/w/index.php?title=Scanning_electron_microscope&oldid=1100146463

Wu, L., Qiu, L., Du, Y., Zeng, F., Lu, Q., Tan, Z., Yin, L., Chen, L., & Zhu, J. (2021). Structure and mechanical properties of PVD and CVD TiAlSiN coatings deposited on cemented carbide. *Crystals*, *11*(6), 1–12. https://doi.org/10.3390/cryst11060598

Xie, W., Zhao, Y., Liao, B., Wang, S., & Zhang, S. (2022). Comparative tribological behavior of TiN monolayer and Ti/TiN multilayers on AZ31 magnesium alloys. *Surface and Coatings Technology*, *441*(March), 128590. https://doi.org/10.1016/j.surfcoat.2022.128590

Zaharah, A. M., Fazira, M. F., Talib, R. J., Mahaidin, A. A., & Selamat, M. A. (2014). Friction and wear characteristics of WC and TiCN-coated insert in turning carbon steel workpiece. *Procedia Engineering*, *68*, 716–722. https://doi.org/10.1016/j.proeng.2013.12.244

7 Transmission Electron Microscope (TEM)

Transmission electron microscope (TEM) is another microscope that uses electron microscopy techniques like scanning electron microscope (SEM) or field emission scanning electron microscope (FESEM) for the generation of images. Also, the images captured using TEM are black and white, similar to SEM and FESEM. When considering the interaction of electrons with the sample, unlike SEM and FESEM which detect and generate images from reflected or knocked-off electrons, TEM captures the transmitted electrons through the sample, giving the technology the name transmission electron microscopy. These transmitted electrons are then magnified and focused on a screen. The screen can be a sensor like a scintillator, a fluorescent screen, or a photographic film. The TEMs are capable of extremely high magnifications going up to 2 million times. The level of magnification achieved through a TEM can be imagined by the fact that a TEM is capable of taking an image of a single column of atoms. These high magnification levels make TEM an excellent choice for analytical research in various fields of science and engineering, such as physical, chemical, and biological sciences. The major application of TEM can be seen in material science, virology, nanotechnology, and cancer research (Britannica, n.d.; Cheadle Center for Biodiversity and Ecological Restoration, n.d.; Stoyanov, 1968; Wikipedia contributors, 2022b). Although, TEM has many advantages over SEM and FESEM, the process pays special attention to sample preparation. As the TEM generates images from the transmitted electrons, the samples under testing should be very thin. The thickness of the sample should be around 100 nm and also they should be readily polished to maintain the same surface quality throughout. Thus, the sample preparation for TEM is a tedious and time-consuming process when compared to SEM and FESEM. However, the transmission capability of electrons can be increased by using a high voltage transmission electron microscope. A maximum of 1 μm sample can be investigated using TEM (Britannica, n.d.; Cheadle Center for Biodiversity and Ecological Restoration, n.d.). In this regard, the present chapter would elaborate construction, working, sample preparation, and application of TEM in different fields with examples.

DOI: 10.1201/9781003340546-8

7.1 CONSTRUCTION

Figure 7.1 shows the Morgagni 268D transmission electron microscope. A general schematic representation of TEM is shown in Figure 7.2. TEM consists of an electron gun for the generation of electrons. The construction and working principle of the electron gun has already been discussed in Sections 5.1 and 6.1. The electron gun is connected to a high-voltage supply of 100 to 300 kV, which provides sufficient energy for the generation of electrons to the electron gun. The emission of electrons can be due to a thermionic effect if a hairpin-shaped tungsten filament is used or may be due to field emission if a field emission gun is used. TEM employs an electromagnetic lens system to focus and condense the electron beam onto the sample. The electron lenses employed in TEM use electromagnetic coils to generate convex lenses. The use of electromagnetic lens arrangement gives TEM the capability to provide focusing up to atomic levels and also magnify this atomic level image on a camera screen (Britannica, n.d.; Cheadle Center for Biodiversity and Ecological Restoration, n.d.; Stoyanov, 1968; Wikipedia contributors, 2022b).

The detectors play a significant role in TEM. The image or diffraction pattern obtained from TEM can be observed using a traditional fluorescent screen which is made from ZnS or ZnS/CdS powder. The excitation of this powder due to the falling electron beam results in the development of an image or diffraction pattern via cathodoluminescence. The image formation takes place due to the emission of

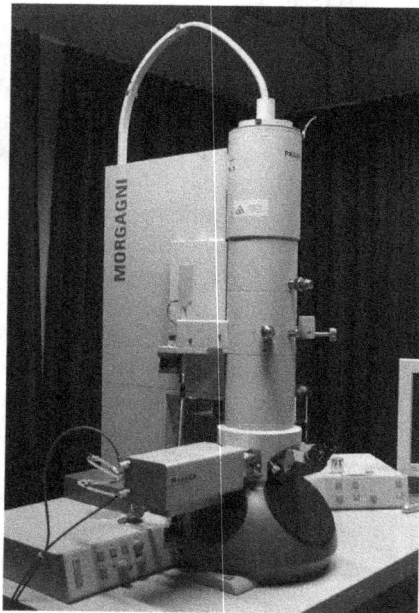

FIGURE 7.1 Transmission electron microscope (Morgagni 268D). (https://commons.wikime dia.org/wiki/File:Transmission_electron_microscope_(Morgagni_268D)_pl.jpg. This file is licensed under the Creative Commons Attribution-Share Alike 3.0 Unported, 2.5 Generic, 2.0 Generic, and 1.0 Generic license.)

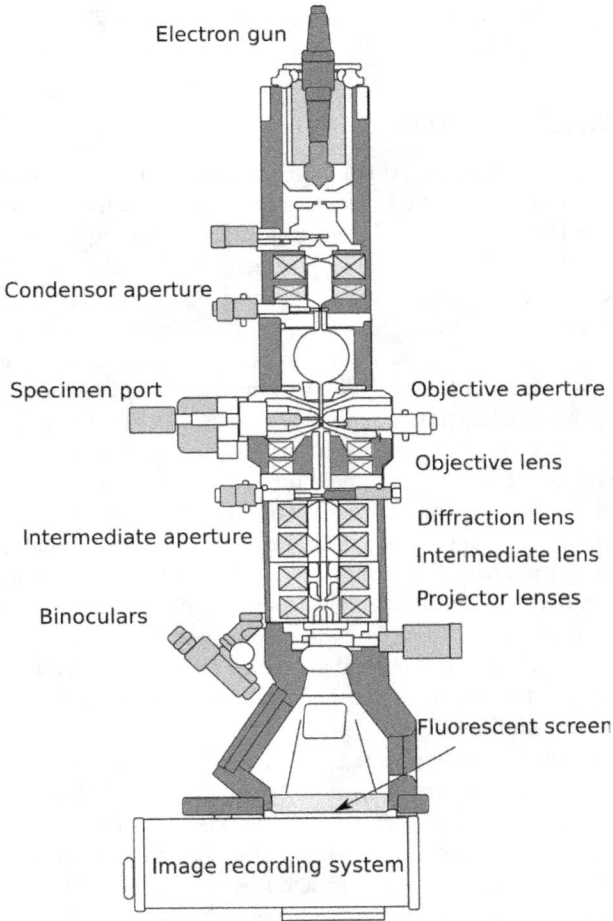

FIGURE 7.2   Schematic representation of transmission electron microscope (TEM).

photons with wavelengths in the visual spectrum. These images can then be developed on a photographic film. A charge-coupled device (CCD) camera is also an important detector for capturing images from TEM. CCDs are usually coupled with a scintillator material sensor made of materials like yttrium aluminum garnet (YAG) that exhibits luminescence when excited by ionizing radiation. Complementary metal oxide semiconductor cameras (CMOS) can also be used for capturing an image in TEM. However, modern TEMs implement more advanced direct electron detectors like hybrid pix that have high detective quantum efficiency (DQE). These detectors can be directly exposed to the electron beam without converting them to photons. The TEM sample stage is also a very significant component that should have high resistance to mechanical drift. A TEM specimen stage usually has airlocks so that the sample holder insertion can be facilitated without vacuum loss. The stage is moved to focus the electron beam on the sample surface that is of significance (Cheadle Center for Biodiversity and Ecological Restoration, n.d.; Professor Zhong L. Wang's Nano Research Lab, n.d.; Stoyanov, 1968; Wikipedia contributors, 2022a).

7.2 WORKING PRINCIPLE

As discussed earlier, the image formation in TEM takes place by collecting the information carried by the transmitted electrons. However, the thickness of the sample should be appropriate so that sufficient electrons transmit through the sample, resulting in a clear and crisp image on the viewing screen. Further, higher resolution images in TEM can be obtained by employing very thin samples and high energies. As far as operation modes are considered, TEM operates in two basic modes, namely, imaging mode and diffraction mode. Figure 7.3 shows both the imaging and diffraction modes of operation on TEM. In both modes, the specimen under investigation is struck by a parallel electron beam which is condensed by a condenser lens arrangement. After the electron beam strikes the specimen, the electrons can transmit in two different ways. Firstly, there will be electrons that remain unscattered and don't change their course of motion, and secondly, there will be electrons that will scatter and change their path of movement due to interaction with the specimen or sample. The unscattered electrons would form a bright central beam on the diffraction pattern. A diffraction pattern of the scattered and diffracted electrons is generated by the objective lens at the back focal plane (BFP) which is then combined to form an image at the image plane (IP). The formed image is termed an intermediate image. This proves the simultaneous presence of both image and diffraction pattern in a transmission electron microscopic examination.

In the imaging mode, the objective aperture is placed in the BFP of the objective lens. This BPF is a place where diffraction spots are formed. If the objective aperture is adjusted to allow only the central beam to pass, then the image formed would be affected by the unscattered electrons and thus the image is called a bright field (BF) image. On the contrary, if the diffracted beam is allowed to pass through the objective aperture, a dark field (DF) image would be formed. If both the diffracted and central beam signals are allowed to pass through the objective aperture, a high-resolution (HR) image is generated. The intermediate and projector lenses help in magnifying and projecting the image onto the camera screen. In the diffraction mode of TEM, the objective aperture is placed in the IP of the objective lens. A variation of current strength to the intermediate lenses helps in projecting the diffraction pattern on the display screen. The diffraction patterns are very helpful in determining the crystal orientation and cell reconstruction investigations (Stoyanov, 1968; Wikipedia contributors, 2022b). This is how the images are formed on a TEM. The basic requirement of TEM is that the samples should be very thin, up to 100 nm thickness. Thus, it is necessary to understand the sample preparation process for TEM more precisely.

7.3 SAMPLE PREPARATION

In TEM sample preparation is a tedious and the most significant part of transmission electron microscopic examinations (Figure 7.4). As the process of image formation in TEM is dependent on the transmitted electrons, the thickness of samples should be less than 100 nm. The sample preparation for TEM is highly specific to the material under

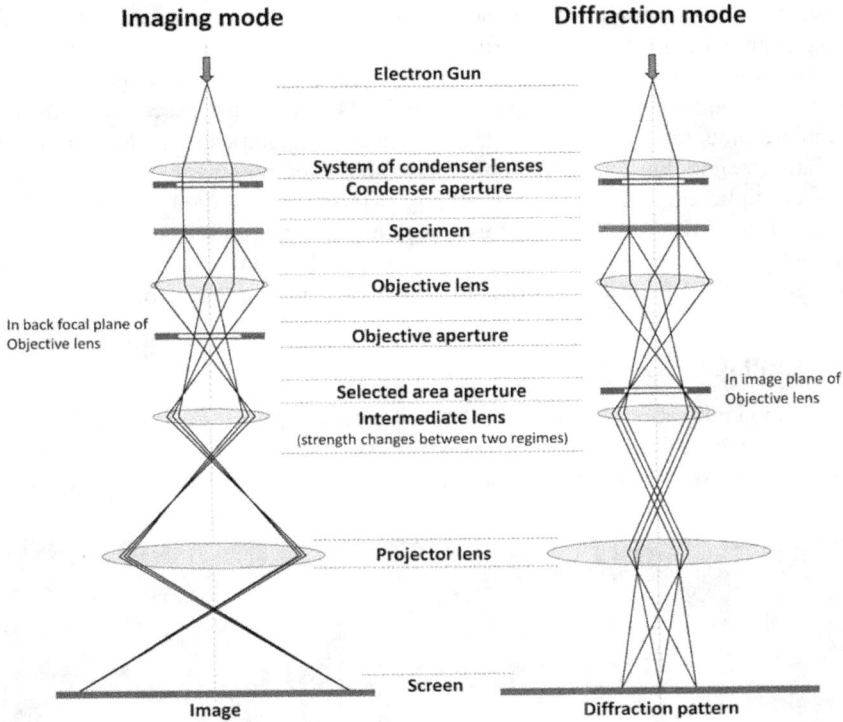

FIGURE 7.3 Schematic representation of image and diffraction modes of operation on a transmission electron microscope.

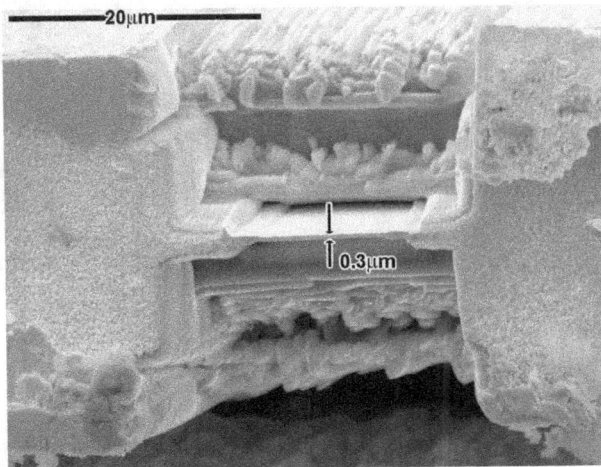

FIGURE 7.4 SEM micrograph showing TEM sample prepared by focused ion beam (FIB) milling. (https://commons.wikimedia.org/wiki/File:Fib_tem_sample.jpg. This file is licensed under the Creative Commons Attribution-Share Alike 3.0 Unported license.)

investigation and would also depend on the information required from the microscopic examination. The two important control parameters of sample preparation for investigations on TEM are the thickness and surface quality of the sample. There are various techniques of sample preparation for TEM, such as tissue sectioning using an ultramicrotome, mechanical milling, chemical etching, ion etching, and ion milling. The latest technology for sample preparation that is predominantly used in different fields of application of TEM is ion milling, which is also called focused ion beam (FIB) milling. FIB is a process that uses high-energy gallium ions for cutting desired size and shape of samples from large specimens (Gharam et al., 2010; Huseynov et al., 2016; McMullan, 1995; Stoyanov, 1968; Wikipedia contributors, 2022b).

7.4 APPLICATION

The high magnification levels make TEM an excellent choice for analytical research in various fields of science and engineering, such as physical, chemical, and biological sciences. The major application of TEM can be seen in material science,

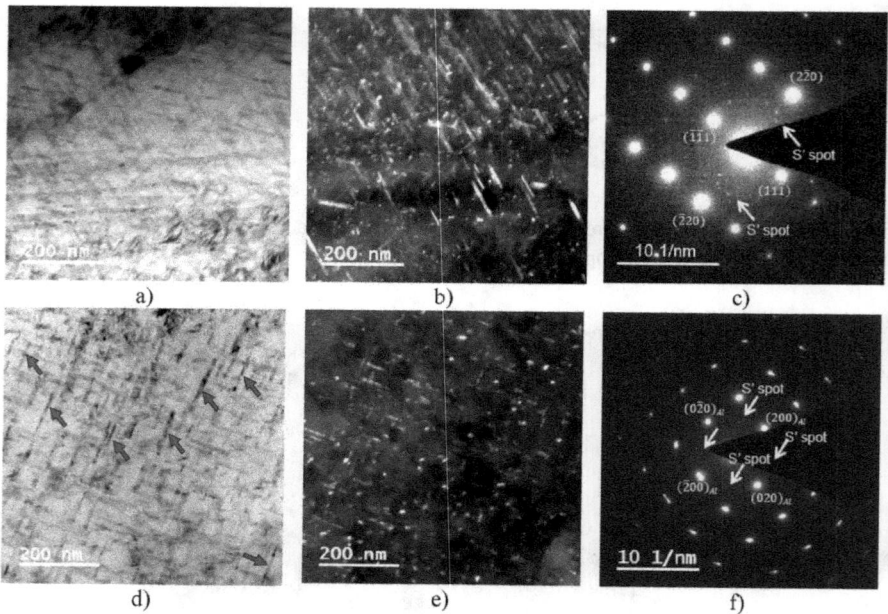

FIGURE 7.5 Different TEM images showing over-aging conditions in AA2024 alloy: (a) bright-field (BF) image, (b) dark-field (DF) image, (c) selected area diffraction (SAD) pattern showing S' precipitates in 50% cold rolled sample at 200 °C for 4 hours, whereas (d) bright field (BF) image, (e) dark field (DF) image and (f) selected area diffraction (SAD) pattern showing S' precipitates in 50% cold rolled sample at 220 °C for 6 hours. (For details, see Mousavi Anijdan, S. H., Sadeghi-Nezhad, D., Lee, H., Shin, W., Park, N., Nayyeri, M. J., Jafarian, H. R., & Eivani, A. R. (2021). TEM study of S' hardening precipitates in the cold rolled and aged AA2024 aluminum alloy: influence on the microstructural evolution, tensile properties & electrical conductivity. *Journal of Materials Research and Technology*, 13, 798–807. https://doi.org/10.1016/j.jmrt.2021.05.003. Reprinted with permission from Elsevier.)

virology, nanotechnology, and cancer research (Britannica, n.d.; Cheadle Center for Biodiversity and Ecological Restoration, n.d.; Stoyanov, 1968; Wikipedia contributors, 2022b). Some of the crucial applications of TEM are analysis of micro-structural surface morphology (Gharam et al., 2011; Huseynov et al., 2016; Mousavi Anijdan et al., 2021), cross-sectional morphology of thin films (Miletić et al., 2014; Wu et al., 2021), phase precipitation in materials (Zhao et al., 2001), crystal structure and orientation (Chang & Duh, 2016; Fukumoto et al., 2009; Koseki et al., 2017), and so on.

Figure 7.5 shows an example of TEM where the transmission electron microscopic technology has been used to study the over-aging conditions in AA2024 alloy with

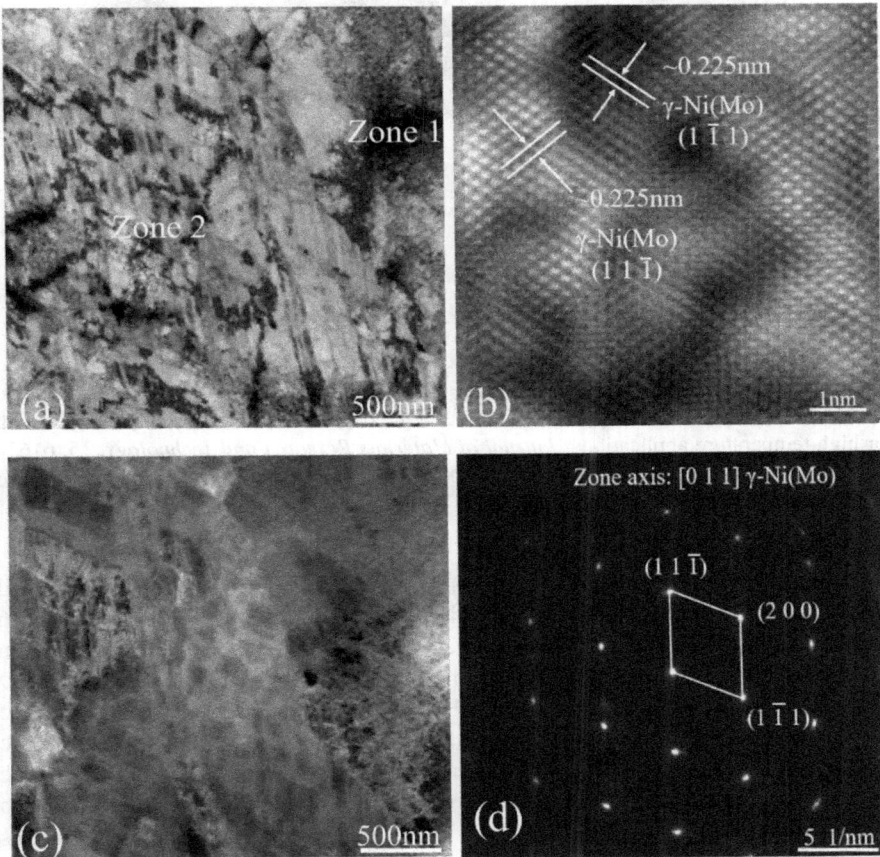

FIGURE 7.6 Surface morphology TEM images of nickel-molybdenum (Ni-Mo) coating: (a) bright-field image, (b) HRTEM image, (c) dark-field image, and (d) fast Fourier transform (FFT) of (b). (For details, see Yang, K., Chen, C., Xu, G., Jiang, Z., Zhang, S., & Liu, X. (2022). HVOF sprayed Ni–Mo coatings improved by annealing treatment: microstructure characterization, corrosion resistance to HCl and corrosion mechanisms. *Journal of Materials Research and Technology*, 19, 1906–1921. https://doi.org/10.1016/j.jmrt.2022.05.181. Reprinted with permission from Elsevier.)

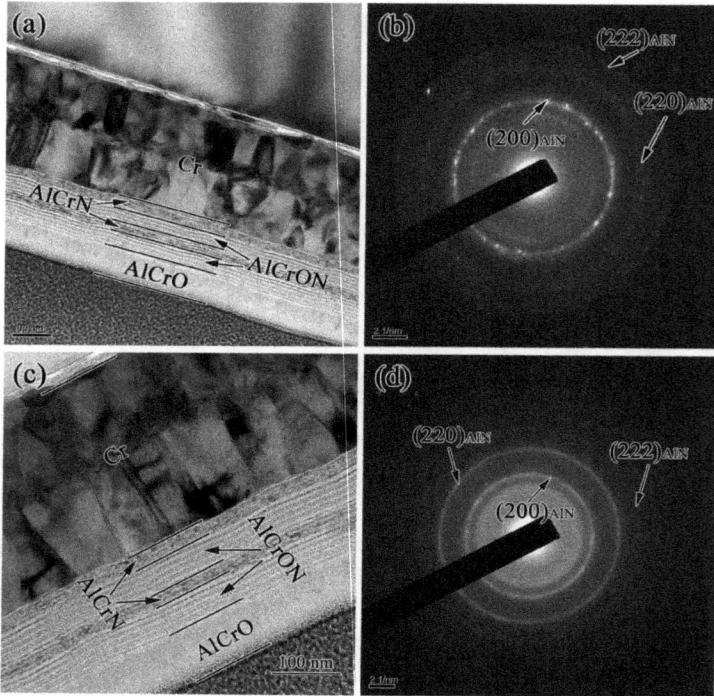

FIGURE 7.7 TEM images showing cross-sectional images of AlCrON-based multilayered coating. (For details, see Wang, X., Yuan, X., Gong, D., Cheng, X., & Li, K. (2021). Optical properties and thermal stability of AlCrON-based multilayer solar selective absorbing coating for high temperature applications. *Journal of Materials Research and Technology*, 15, 6162–6174. https://doi.org/10.1016/j.jmrt.2021.11.068. This is an open access article distributed under the terms of the Creative Commons CC-BY license by Elsevier.)

the help of bright field (BF) TEM images, dark field (DF) TEM images, and selected area diffraction (SAD) patterns (Mousavi Anijdan et al., 2021). Further, Figure 7.6 illustrates another example where the surface morphology of the nickel-molybdenum (Ni-Mo) coating has been studied with the help of BF and DF TEM images, SAD patterns, and Field Fourier Transform (FFT) (Yang et al., 2022). Also, TEM can be a useful tool for analyzing the cross-sectional morphology of thin films that have multilayers at a nanoscale. One such example has been illustrated in Figure 7.7 where the cross-sectional morphology of AlCrON-based multilayered coating has been studied using TEM (Wang et al., 2021). These are some of the common examples where TEM has crucial applications.

7.5 SUMMARY

TEM is another microscope using electron microscopy techniques like SEM and FESEM for generating images. When considering the interaction of electrons with the sample, unlike SEM and FESEM which detect and generate images from reflected or knocked-off electrons, TEM captures the transmitted electrons through the sample, giving the technology the name transmission electron microscopy. The major applications of TEM can be seen in material science, virology, nanotechnology, and cancer research. Some of the crucial TEM research implementations such as investigating cross-sectional morphology of thin-films, and surface morphology of different materials has been discussed. Although TEM has various advantages over SEM and FESEM techniques, the specimen preparation is quite tedious task that requires preparation of very thin samples having thickness less than 100 nm for conventional TEM. Unlike other imaging techniques, TEM can get diffraction patterns of the sample, which makes it very effective in studying crystal orientation and cell reconstruction. These capabilities of TEM make it a significant analytical tool in different fields of science and engineering.

REFERENCES

Britannica. (n.d.). Retrieved August 16, 2022, from www.britannica.com/technology/transmission-electron-microscope.

Chang, C. C., & Duh, J. G. (2016). Duplex coating technique to improve the adhesion and tribological properties of CrAlSiN nanocomposite coating. *Surface and Coatings Technology*. https://doi.org/10.1016/j.surfcoat.2016.11.032

Cheadle Center for Biodiversity and Ecological Restoration. (n.d.). Retrieved August 16, 2022, from www.ccber.ucsb.edu/ucsb-natural-history-collections-botanical-plant-anatomy/transmission-electron-microscope#:~:text=Transmission electron microscopes (TEM) are,how small a cell is

Fukumoto, N., Ezura, H., & Suzuki, T. (2009). Synthesis and oxidation resistance of TiAlSiN and multilayer TiAlSiN/CrAlN coating. *Surface and Coatings Technology*, *204*(6–7), 902–906. https://doi.org/10.1016/j.surfcoat.2009.04.027

Gharam, A. A., Lukitsch, M. J., Balogh, M. P., & Alpas, A. T. (2010). High temperature tribological behaviour of carbon based (B4C and DLC) coatings in sliding contact with aluminum. *Thin Solid Films*, *519*(5), 1611–1617. https://doi.org/10.1016/j.tsf.2010.07.074

Gharam, A. A., Lukitsch, M. J., Balogh, M. P., Irish, N., & Alpas, A. T. (2011). High temperature tribological behavior of W-DLC against aluminum. *Surface and Coatings Technology*, *206*(7), 1905–1912. https://doi.org/10.1016/j.surfcoat.2011.08.002

Huseynov, E., Garibov, A., & Mehdiyeva, R. (2016). TEM and SEM study of nano SiO_2 particles exposed to influence of neutron flux. *Journal of Materials Research and Technology*, *5*(3), 213–218. https://doi.org/10.1016/j.jmrt.2015.11.001

Koseki, S., Inoue, K., Sekiya, K., Morito, S., Ohba, T., & Usuki, H. (2017). Wear mechanisms of PVD-coated cutting tools during continuous turning of Ti-6Al-4V alloy. *Precision Engineering*, *47*, 434–444. https://doi.org/10.1016/j.precisioneng.2016.09.018

McMullan, D. (1995). Scanning electron microscopy 1928–1965. *Scanning*, *17*, 175–185.

Miletić, A., Panjan, P., Škorić, B., Čekada, M., Dražič, G., & Kovač, J. (2014). Microstructure and mechanical properties of nanostructured Ti-Al-Si-N coatings deposited by magnetron sputtering. *Surface and Coatings Technology*, *241*, 105–111. https://doi.org/10.1016/j.surfcoat.2013.10.050

Mousavi Anijdan, S. H., Sadeghi-Nezhad, D., Lee, H., Shin, W., Park, N., Nayyeri, M. J., Jafarian, H. R., & Eivani, A. R. (2021). TEM study of S' hardening precipitates in the cold rolled and aged AA2024 aluminum alloy: influence on the microstructural evolution, tensile properties & electrical conductivity. *Journal of Materials Research and Technology*, *13*, 798–807. https://doi.org/10.1016/j.jmrt.2021.05.003

Professor Zhong L. Wang's Nano Research Lab. (n.d.). *Fundamental Theory of Transmission Electronic Microscopy*. https://nanoscience.gatech.edu/zlwang/research/tem.html

Stoyanov, P. A. (1968). *Transmission Electron Microscopes*. Springer, ISBN 978-0-387-76500-6 hardcover ISBN 978-0-387-76502-0 softcover (This is a four-volume set. The volumes are not sold individually.) e-ISBN 978-0-387-76501-3

Wang, X., Yuan, X., Gong, D., Cheng, X., & Li, K. (2021). Optical properties and thermal stability of AlCrON-based multilayer solar selective absorbing coating for high temperature applications. *Journal of Materials Research and Technology*, *15*, 6162–6174. https://doi.org/10.1016/j.jmrt.2021.11.068

Wikipedia contributors. (2022a). *Detectors for transmission electron microscopy – Wikipedia, The Free Encyclopedia*. https://en.wikipedia.org/w/index.php?title=Detectors_for_transmission_electron_microscopy&oldid=1104552096

Wikipedia contributors. (2022b). *Transmission electron microscopy – Wikipedia, The Free Encyclopedia*. https://en.wikipedia.org/w/index.php?title=Transmission_electron_microscopy&oldid=1103921398

Wu, L., Qiu, L., Du, Y., Zeng, F., Lu, Q., Tan, Z., Yin, L., Chen, L., & Zhu, J. (2021). Structure and mechanical properties of PVD and CVD TiAlSiN coatings deposited on cemented carbide. *Crystals*, *11*(6), 1–12. https://doi.org/10.3390/cryst11060598

Yang, K., Chen, C., Xu, G., Jiang, Z., Zhang, S., & Liu, X. (2022). HVOF sprayed Ni–Mo coatings improved by annealing treatment: microstructure characterization, corrosion resistance to HCl and corrosion mechanisms. *Journal of Materials Research and Technology*, *19*, 1906–1921. https://doi.org/10.1016/j.jmrt.2022.05.181

Zhao, J. C., Ravikumar, V., & Beltran, A. M. (2001). Phase precipitation and phase stability in Nimonic 263. *Metallurgical and Materials Transactions A: Physical Metallurgy and Materials Science*, *32*(6), 1271–1282. https://doi.org/10.1007/s11661-001-0217-4

8 Atomic Force Microscope (AFM)

Atomic force microscope (AFM) is a type of scanning probe microscope used to determine properties such as height, magnetic force, surface potential, and friction, and also has the ability to measure intermolecular forces. AFM amplifies the image of the sample and uses a cantilever made of silicon with a low spring constant to image the sample. AFM consists of an optical head, a movable scanner, and a multimode base. The head contains the probe, laser, photodiode array, and adjustment knobs used to align the system. The sample is placed on the scanner, which contains the piezo tube that controls the movement of the sample. The base controls the raising and lowering of the probe. A laser beam is constantly reflected from the top of the cantilever. The beam detects the bending of the boom and calculates the actual position of the boom. The AFM records a three-dimensional image of the external topography of the sample under a constant force, resulting in an image with maximum resolution. The Van der Waal forces, capillary forces, adhesion forces, and double layer forces balance the interaction between the tip of the cantilever and the sample surface. The operating modes of the AFM are contact mode, non-contact mode, and capture mode. In contact mode, the cantilever tip scans laterally across the sample surface. The tapping mode provides high-resolution topographic imaging of subcellular structures of the sample. These technologies are also useful in chemical sciences, nanotechnology, and single molecule experiments. However, AFM is a non-element-specific measurement technique. The AFM can only reach a lower size limit of 0.1 nm in the Z-direction (image height). For use in a harsh environment, the tips can be coated with either platinum, iridium, or gold. AFM probes can be manufactured with different force constants and resonant frequencies for the three imaging modes: Contact mode, NC/Tapping mode, and force modulation mode measurements. Harsh environmental conditions include applications in a variety of wet as well as high radiation environments. The gold-coated tip is of particular interest for biology and life science applications. This chapter covers the operation, setup, and application of the AFM.

8.1 CONSTRUCTION

The AFM combines performance and a wide range of applications important to materials science research. Its unique direct drive tip scanner technology combined

DOI: 10.1201/9781003340546-9

with Clean Drive is the key to fast and stable operation in air and liquid. The tip scanner design makes the AFM's performance independent of the mass of the sample under study, allowing measurements on heavy samples. Full motorization not only simplifies working with the system but also enables automated measurements involving different areas of a sample. In addition to reliable topographic imaging, the AFM also offers a whole range of different modes for studying nanoelectrics. The universe of accessories available for the AFM offers advanced features such as heating or cooling the sample, applying a variable magnetic field, sensing low electric currents, or studying the changes that occur during electrochemical processes at electrodes with in situ AFM imaging (Reents et al., 1860). True atomic resolution of silicon surface 7x7 Atomic images of this surface obtained by STM had convinced the scientific community of the spectacular spatial resolution of scanning tunneling microscopy but had to wait a little longer to be used by. In manipulation, the forces between the tip and the sample can also be used to change the properties of the sample in a controlled manner. Examples include manipulation of atoms, scanning probe lithography, and local stimulation of cells (Kawakatsu et al., 2006).

Simultaneously with the acquisition of topographic images, other properties of the sample can be measured locally and displayed as an image, often at similarly high resolution. Examples of such properties include mechanical properties (such as stiffness or adhesion) and electrical properties (such as conductivity or surface potential). Most scanning probe microscopy techniques are extensions of AFM that use this modality (Galvanetto, 2018)

The main difference between AFM and competing technologies such as light microscopy and electron microscopy is that AFM does not use lenses or beams. Therefore, spatial resolution is not limited by diffraction and aberration, and preparation of space for beam delivery (by creating a vacuum) and staining of the sample is not required. AFM allows the determination of surface topography in scanning probe mode (Lapshin & Obyed kov, 1993). There are different versions of AFM, such as "Constant Height Mode," "Constant Force Mode," "Force Modulation Microscopy," "Phase Imaging Mode," "Kelvin Probe Force Microscopy," and "Scanning Capacitance Microscopy," which allow obtaining the results. The basic elements of AFM include a scanning element, a feedback system, a detector of deflection and movement of the cantilever (laser + photodiode), a piezo scanner, and a vibration guard (Figure 8.1). The interaction between the sharp probe and the sample is represented by the Van der Waals forces (the forces of intermolecular interaction with an energy of 10–20 kJ/mol). The interaction energy of two atoms can be approximated by the Lennard–Jones potential (Wang et al., 2020). Except in the equilibrium position, a force (attractive or repulsive) acts on the cantilever, causing the bending of an elastic cantilever, which is registered by a deflection and motion detector consisting of a photodiode and a laser. The laser radiation is focused on the cantilever, then reflected and falls on the photodiode. The semiconductor photodetector consists of four parts. When there is zero deformation of the bracket, the reflected beam enters directly into the center of the photosensitive area, while any other deformation changes the position of the beam. In this way, all changes in the position of the console in space are detected.

AFM can be used for force spectroscopy, i.e., direct measurement of tip-sample interaction forces as a function of tip-sample distance. The result of this measurement

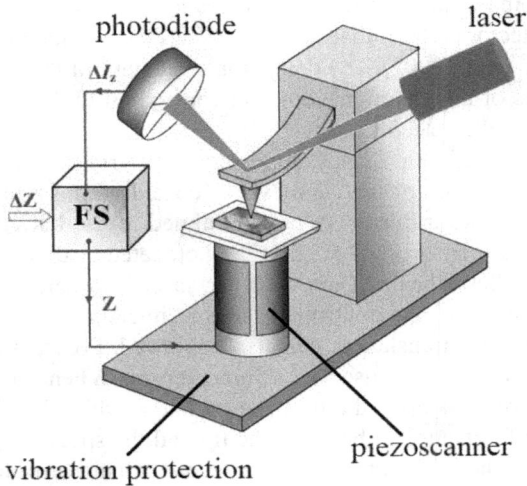

FIGURE 8.1 Schematic representation of atomic force microscope.

is called a force-displacement curve. In this method, the AFM tip is extended to the surface and retracted, while the deflection of the cantilever is monitored as a function of piezoelectric displacement. These measurements have been used to measure nano-scale contacts, atomic bonding, Van der Waals forces and Casimir forces, dissolution forces in liquids, and stretching and breaking forces of single molecules (Hinter dorfer & Dufrêne, 2006). AFM has also been used to measure the dispersion force of polymers adsorbed on the substrate in an aqueous environment (Lapshin & Obyed kov, 1993). Forces in the order of a few piconewtons can now be routinely measured with a vertical distance resolution of better than 0.1 nm. Force spectroscopy can be performed in either static or dynamic modes. In dynamic modes, information about the vibration of the cantilever is acquired in addition to the static deflection (Butt et al., 2005).

Among the problems with this technique is that the distance between the tip and the specimen cannot be measured directly and that low stiffness cantilevers are often required, which tend to "snap" to the surface. However, these problems are not insur-mountable. An AFM has been developed that directly measures the distance between the tip and the sample (King et al., 2009). Snapping can be reduced by measuring in liquids or by using stiffer cantilevers, but in the latter case, a more sensitive deflection sensor is required. By attaching a small dither to the tip, the stiffness (force gradient) of the joint can also be measured (Hoffmann et al., 2001).

8.2   AFM WORKING PRINCIPLE

The AFM principle is based on the cantilever/tip assembly interacting with the sample; this assembly is also commonly referred to as the probe. The AFM probe interacts with the substrate through a raster scanning motion. The up/down and side-to-side motion of the AFM tip as it scans along the surface is monitored by a laser beam reflected from the cantilever. This reflected laser beam is tracked by a position-sensitive photodetector (PSPD) that records the vertical and lateral movement of the probe (Yamaguchi et al., 2007). The deflection sensitivity of these detectors must be calibrated in terms of how many nanometers correspond to the movement of a unit voltage measured at the detector.

An AFM consists of four main components, the first of which is a cantilever with a sharp tip. The back of the cantilever is coated with a reflective material so that it reflects light like a mirror. A laser is aimed at the back of the cantilever, and the reflected light is then collected by a photodetector that is very sensitive to the positions of the laser beam. The cantilever, laser, and detector system are precisely scanned by an electrical controller. The specimen, which is located under the microscope on a DeepL translation, can also be moved. Figure 8.2 shows a typical AFM setup. The cantilever is used as a force sensor. When the tip of the cantilever is scanned over a sample, the tip behaves like a rubber band (Umeda, 1991). Depending on the force applied between the tip and the specimen, the cantilever is compressed or stretched. The photodetector records changes in the position of the reflected laser beam proportional to the movement of the cantilever. By scanning the surface of the material, a detailed topographic image of the sample can be acquired. Various types of forces can be detected between the AFM tip and the sample, including electrostatic forces between charges on the surface of the sample and the tip. Other forces detected include Van der Waals forces, mechanical contact forces, capillary forces, magnetic forces, Casimir forces, and forces caused by chemical bonding. AFM can be operated in different modes to best suit the type of force being measured (Alsteens et al., 2008). In static mode, also called contact mode, the AFM tip is in contact with the sample surface when it is scanned. The tip is "dragged" across the sample like a needle. The tip behaves like an elastic element that contracts and expands when it encounters a force caused by the sample surface. The changes

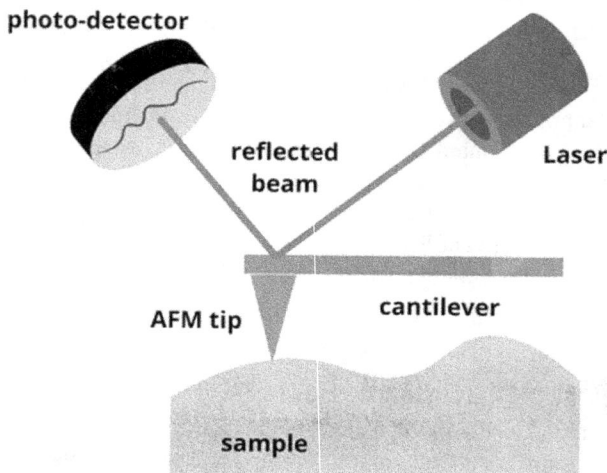

FIGURE 8.2   Schematic illustrations of basic principles of AFM.

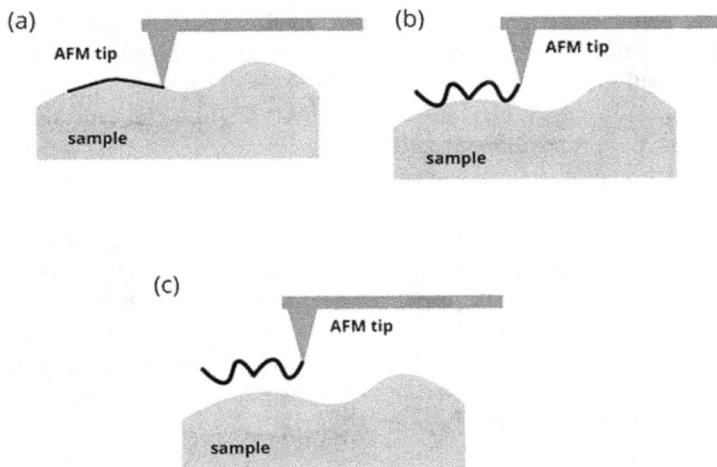

FIGURE 8.3 Schematic illustrations of the operational modes of AFM: (a) Contact mode, (b) tapping mode, and (c) non-contact mode.

in the cantilever deflect the laser beam, which is imaged on the photodetector. Soft materials are further deflected when force is applied, which is why AFM tips are generally made of soft materials. However, because the tip is in contact with the sample over which it is scanning, the sample can be easily damaged. The static mode is most often used for recording hard samples to obtain topographic information, while in the dynamic mode the boom is not "static" but is excited to vibrate at a specific frequency. In dynamic mode, tapping mode and non-contact mode are widely used: In tapping mode, the tip oscillates up and down over the surface of the sample. The tip comes into contact with the sample at its lowest point of oscillation by "tapping" the surface. Repulsive forces are detected when the tip "touches" the surface, while attractive forces are detected at the peak of the oscillation (Zitzler et al., 2002). Tapping generally causes less damage to the surface and tip than the contact method. This technique is used, among other things, to image the formation of molecules in chemistry, where the tip of the cantilever does not touch the sample surface (Rugar et al., 1989).

The cantilever is made to vibrate above the surface of the specimen. Therefore, the tip does not cause damage to the specimen. In the non-contact mode, the drive motion of the cantilever is either amplitude or frequency modulated, and these modulation techniques offer different detection sensitivities depending on the application. The non-contact measurement method is advantageous for imaging soft samples. Some applications include imaging biological samples and organic thin films. Figure 8.3 shows how the AFM works.

Challenges and new developments in AFMs in addition to the damage to samples, AFM has other disadvantages. The size of an AFM image that can be acquired in a single scan is small compared to other scanning microscopy techniques. AFM can be combined with a variety of experimental techniques developed in other scientific endeavors, including optical microscopy and spectroscopy techniques such as fluorescence microscopy or infrared spectroscopy. These combined methods have led to new investigations such as scanning near-field optical microscopy (Binnig et al., n.d.). Nonoptical detection methods such as q-Plus-based AFM with functionalized tips have also been used successfully. Advances in AFMs promise to advance investigations in photovoltaics and energy storage research, polymer science, nanotechnology, and medical research.

AFM can be used to image and manipulate atoms and structures on a variety of surfaces. The atom at the tip of the tip "senses" the individual atoms on the underlying surface as it forms an incipient chemical bond with each atom. Since these chemical interactions only slightly change the tip's vibrational frequency, they can be detected and mapped. This principle has been used to distinguish between silicon, tin, and lead atoms on an alloy surface by comparing these "atomic fingerprints" with values from density functional theory simulations (DFT) (Sugimoto et al., 2007). The trick is to first measure these forces accurately for each type of atom expected in the sample and then compare them to the forces resulting from simulations of DFT. The team found that the tip interacts most strongly with silicon atoms and 24% and 41% less strongly with tin and lead atoms, respectively. In this way, each atom type in the matrix can be identified as the tip is moved across the surface. Figure 8.4 shows the AFM topographic scan of a stainless steel surface. One can see the micro- and nanoscale features of the glass, which represent the roughness of the material. The image space is (x,y,z) = (10 μm × 10 μm × 320 nm) and the average roughness is 0.31 nm. The image generation is a plotting method in which a color image is generated at each x-y coordinate by changing the x-y position of the tip during scanning and recording the measured variable, i.e., the intensity of the control signal (Nishida et al., 2008). The color mapping shows the measured value corresponding to each coordinate. The image expresses the intensity of a value as a hue. Usually, the correspondence between the intensity of a value and a hue is shown in the form of a color scale in the explanations of the image

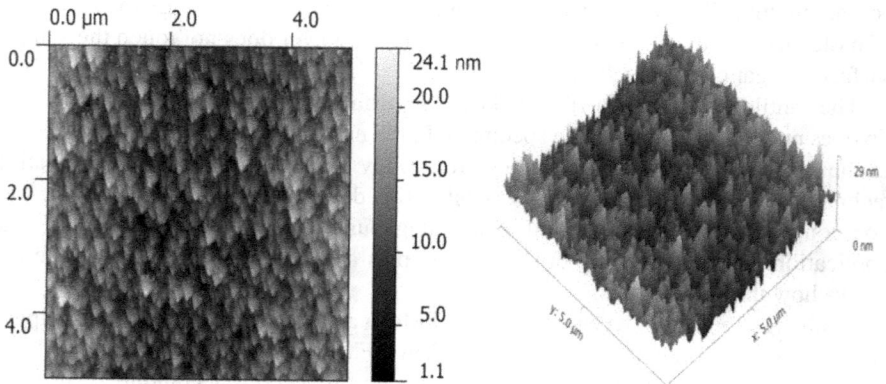

FIGURE 8.4 Atomic force microscope topographical scan of a stainless steel surface.

Roughness Parameters:

Selected line:	(19, 109) to	(0.37, 2.14) to
	(245, 109) px	(4.80, 2.15) μm

Amplitude

Roughness average	**(Ra):**	**0.31 nm**
Root mean square roughness	**(Rq):**	**0.37 nm**
Maximum height of the roughness	**(Rt):**	**1.76 nm**
Maximum roughness valley depth	(Rv):	0.83 nm
Maximum roughness peak height	(Rp):	0.93 nm
Average maximum height of the roughness	(Rtm):	1.32 nm
Average maximum roughness valley depth	(Rvm):	0.65 nm
Average maximum roughness peak height	(Rpm):	0.67 nm
Average third highest peak to third lowest valley height	(R3z):	1.57 nm
Average third highest peak to third lowest valley height	(R3z ISO):	0.95 nm
Average maximum height of the profile	**(Rz):**	**1.60 nm**
Average maximum height of the roughness	(Rz ISO):	1.32 nm
Skewness	(Rsk):	0.007
Kurtosis	(Rku):	2.376
Waviness average	(Wa):	11.11 nm
Root mean square waviness	(Wq):	11.28 nm
Waviness maximum height	(Wy=Wmax):	16.81 nm
Maximum height of the profile	(Pt):	17.25 nm

Spatial

Average wavelength of the profile	(λa):	0.11 μm
Root mean square (RMS) wavelength of the profile	(λq):	878.10 μm

Hybrid

Average absolute slope	(Δa):	17.76 10^-3
Root mean square (RMS) slope	(Δq):	2.658 10^-6
Length	(L):	4.43 μm
Developed profile length	(L0):	4.43 μm
Profile length ratio	(lr):	1.001

8.3 APPLICATION

AFM is one of the most effective imaging techniques used at the nanoscale and subnanoscale levels. This technique has been applied to numerous problems in the natural sciences and can capture a range of material surface properties in both liquid media and air. Disciplines in which AFM is used include semiconductor science and technology, thin film and coatings, tribology (surface and frictional interactions), surface chemistry, polymer chemistry and physics, cell biology, molecular biology, energy storage (battery), and power generation (photovoltaic) materials, piezoelectric and ferroelectric materials (Fukuma, 2010). Applications in the field of solid-state physics include (a) the identification of atoms on a surface, (b) the evaluation of interactions between a given atom and its neighboring atoms, and (c) the study of changes in physical properties resulting from changes in an atomic arrangement due to atomic manipulation. In Molecular Biology (Giessibl, n.d.), AFM can be used to study the structure and mechanical properties of protein complexes and assemblies. In cell biology, AFM can be used to try to distinguish cancer cells from normal cells based on cell hardness and to evaluate the interactions between a particular cell and its neighboring cells in a competitive culture system (Carpick & Salmeron, 1997). AFM can also be used to indent cells to study how they regulate the stiffness or shape of the cell membrane or wall. In some versions, electrical potentials can also be sampled with conducting cantilevers. In more advanced versions, currents can be passed through the tip to study electrical conductivity or transport of the under-lying surface, but this is a challenging task for which few research groups provide consistent data (Roiter & Minko, 2005). AFM is a powerful imaging and measure-ment technique that has become critical for nanoscale research and industrial research and development (R&D) in all its possible forms. It is also used in the imaging and development of graphene, and AFM has enabled the study and characterization of graphene composites. Industries such as aerospace and automotive rely heavily on AFM to develop materials. AFM is so versatile that it can determine a wide range of mechanical properties at the nanoscale and fully characterize a material sample in hours rather than days (Hasselbach et al., 2008). AFM can be used in biological research to distinguish cancer cells from normal cells based on their stiffness. Every day there are new applications for AFM, and the number of research areas is almost unlimited.

8.4   SUMMARY

The ability of AFM to provide high-resolution images and measure small forces in the natural environment of the sample makes this technique invaluable for characterizing materials. Currently, AFM is used in research to determine the function of surface morphology with high-resolution imaging, characterize how nanoscale topography affects material behavior, determine the mechanical properties of materials and their native environment, and understand how physical properties change. AFM can be used alone or in combination with other optical and spectroscopic techniques to pro-vide complementary information, creating an even more powerful tool.

REFERENCES

Alsteens, D., Verbelen, C., Dague, E., Raze, D., Baulard, A. R., & Dufrêne, Y. F. (2008). Organization of the mycobacterial cell wall: A nanoscale view. *PflugersArchiv European Journal of Physiology*, *456*(1), 117–125. https://doi.org/10.1007/s00424-007-0386-0
Binnig, G., Quate' ', C. F., Gi, E. L., & Gerber, C. (n.d.). *Atomic Force Microscope. Physical Review Letters*, *56*, 930.

Butt, H. J., Cappella, B., & Kappl, M. (2005). Force measurements with the atomic force microscope: Technique, interpretation and applications. *Surface Science Reports, 59*(1–6), 1–152. https://doi.org/10.1016/j.surfrep.2005.08.003

Carpick, R. W., & Salmeron, M. (1997). Scratching the surface: fundamental investigations of tribology with atomic force microscopy. *Chem Rev, 97*(4), 1163–1194. doi: 10.1021/cr960068q

Fukuma, T. (2010). Water distribution at solid/liquid interfaces visualized by frequency modulation atomic force microscopy. *Science and Technology of Advanced Materials, 11*(3), 033003. https://doi.org/10.1088/1468-6996/11/3/033003

Galvanetto, N. (2018). Single-cell unroofing: Probing topology and nanomechanics of native membranes. *Biochimica et Biophysica Acta–Biomembranes, 1860*(12), 2532–2538. https://doi.org/10.1016/j.bbamem.2018.09.019

Giessibl, F. J. (n.d.). *Advances in atomic force microscopy.*

Hasselbach, K., Ladam, C., Dolocan, V. O., Hykel, D., Crozes, T., Schuster, K., & Mailly, D. (2008). High resolution magnetic imaging: MicroSQUID force microscopy. *Journal of Physics: Conference Series, 97*(1). https://doi.org/10.1088/1742-6596/97/1/012330

Hinterdorfer, P., & Dufrêne, Y. F. (2006). Detection and localization of single molecular recognition events using atomic force microscopy. *Nature Methods, 3*(5), 347–355. https://doi.org/10.1038/nmeth871

Hoffmann, P. M., Oral, A., Grimble, R. A., ÖzgürÖzer, H., Jeffery, S., & Pethica, J. B. (2001). Direct measurement of interatomic force gradients using an ultra-low-amplitude atomic force microscope. *Proceedings of the Royal Society A: Mathematical, Physical and Engineering Sciences, 457*(2009), 1161–1174. https://doi.org/10.1098/rspa.2000.0713

Kawakatsu, H., Kawai, S., Kobayashi, D., Kitamura, S. I., & Meguro, S. (2006). Atomic force microscopy utilizing subAngstrom cantilever amplitudes. *e-Journal of Surface Science and Nanotechnology, 4*, 110–114. https://doi.org/10.1380/ejssnt.2006.110

King, G. M., Carter, A. R., Churnside, A. B., Eberle, L. S., & Perkins, T. T. (2009). Ultrastable atomic force microscopy: Atomic-scale stability and registration in ambient conditions. *Nano Letters, 9*(4), 1451–1456. https://doi.org/10.1021/nl803298q

Lapshin, R. v., & Obyedkov, O. v. (1993). Fast-acting piezoactuator and digital feedback loop for scanning tunneling microscopes. *Review of Scientific Instruments, 64*(10), 2883–2887. https://doi.org/10.1063/1.1144377

Nishida, S., Kobayashi, D., Sakurada, T., Nakazawa, T., Hoshi, Y., & Kawakatsu, H. (2008). Photothermal excitation and laser Doppler velocimetry of higher cantilever vibration modes for dynamic atomic force microscopy in liquid. *Review of Scientific Instruments, 79*(12). https://doi.org/10.1063/1.3040500

Reents, W. D., Stroble, F., Freas, R. B., Wronka, J., Ridge, D. P., Loh, S. K., Fisher, E. R., Lian, L., Schultz, R. H., Hsu, -t, Kemper, P. R., Bowers, M. T., Chem, J., Ohnesorge, F., & Binnig, G. (1860). Atomic Force Microscope *International Journal of Mass Spectrometery and Ion Processes, 20*(11). http://science.sciencemag.org/

Roiter, Y., & Minko, S. (2005). AFM single molecule experiments at the solid-liquid interface: In situ conformation of adsorbed flexible polyelectrolyte chains. *Journal of the American Chemical Society, 127*(45), 15688–15689. https://doi.org/10.1021/ja0558239

Rugar, D., Mamin, H. J., & Guethner, P. (1989). Improved fiber-optic interferometer for atomic force microscopy. *Applied Physics Letters, 55*(25), 2588–2590. https://doi.org/10.1063/1.101987

Sugimoto, Y., Pou, P., Abe, M., Jelinek, P., Pérez, R., Morita, S., & Custance, Ó. (2007). Chemical identification of individual surface atoms by atomic force microscopy. *Nature, 446*(7131), 64–67. https://doi.org/10.1038/nature05530

Umeda, N. (1991). Scanning attractive force microscope using photothermal vibration. *Journal of Vacuum Science & Technology B: Microelectronics and Nanometer Structures*, *9*(2), 1318. https://doi.org/10.1116/1.585187

Wang, X., Ramírez-Hinestrosa, S., Dobnikar, J., & Frenkel, D. (2020). The Lennard-Jones potential: When (not) to use it. *Physical Chemistry Chemical Physics*, *22*(19), 10624–10633. https://doi.org/10.1039/c9cp05445f

Yamaguchi, R. T., Miyamoto, K. I., Ishibashi, K. I., Hirano, A., Said, S. M., Kimura, Y., & Niwano, M. (2007). DNA hybridization detection by porous silicon-based DNA micro-array in conjugation with infrared microspectroscopy. *Journal of Applied Physics*, *102*(1). https://doi.org/10.1063/1.2751415

Zitzler, L., Herminghaus, S., & Mugele, F. (2002). Capillary forces in tapping mode atomic force microscopy. *Physical Review B–Condensed Matter and Materials Physics*, *66*(15), 1–8. https://doi.org/10.1103/PhysRevB.66.155436

9 Near-Field Scanning Optical Microscope Raman

9.1 CONSTRUCTION

Early in the 20th century, Raman spectroscopy was developed as a material characterization technique. According to theoretical predictions, photons, similar to X-rays, are predicted to be inelastically scattered from molecules when they hit them, just the way X-rays are scattered by Compton scattering. In 1928, Raman and Krishnan presented the first experimental evidence of such scattering events for the first time (Raman & Krishnan, 1928). There is an obvious difference in energy between the scattered radiation coming from the sample surface and the monochromatic light that is illuminated the sample surface. A well-known example of this effect is the Raman effect, which is the result of inelastic scattering, also known as Raman scattering, and that can be directly attributed to spectroscopic techniques used to study normal modes of vibrational motion in molecules in order to investigate their structural and chemical properties (Krishna et al., 2016).

As an approach to overcome the diffraction limitation, near-field scanning optical microscopy Raman has been proposed using an optical probe with a small aperture that has a resolution less than the wavelength of the probe in order to eliminate the diffraction problem. The combination of optical spectroscopy with scanning probe microscopy has recently been proposed as a general concept for enhancing the spatial resolution of optical imaging beyond the limits of diffraction. There have been a variety of technologies developed to enable near-field optical probing with a small aperture on the metal-coated probe or fiber probe in order to make this possible (Trautman et al., 1994). A complementary approach that uses apertureless near-field optics to achieve near-field imaging was also proposed and the concept of enhanced Raman spectroscopy is derived from a surface enhanced Raman spectroscopy approach (Krishna & Colak, 2022). It has been known since 1985 that surface-enhanced Raman spectroscopy and near-field scanning optical microscopy have been advanced by using scanning tunneling microscopy or atomic force microscopy techniques in conjunction with new generation technologies. Numerous researchers have reported a wide range of studies involving various types of near-field optical microscopes in the past few years (Dong et al., 2018).

One of the major technical challenges in Raman spectroscopy is to detect too weak Raman scattering signals in order to increase the spatial resolution. A tip-enhanced

DOI: 10.1201/9781003340546-10

Raman spectroscopy has been proposed as a method for improving the signal in Raman spectroscopy (Hartschuh et al., 2003). It was discovered later on that there are various types of tip-enhanced near-field Raman spectroscopy developed (Atkin et al., 2012; Lee et al., 2021). The concepts are based on the surface-enhanced Raman spectroscopy principle, whereby the signal is enhanced near the metal particle or tip in order to enhance the signal quality. A Raman scattering near-field probe can be designed by using a metal particle or a metal tip as a Raman scattering near-field probe with a larger amplitude of Raman scattering. In spite of the fact that tip-enhanced Raman Spectroscopy technology has a high spatial resolution based on the enhancement of Raman scattering signals around the metal particle, the metal particle or metal tip presents a serious problem since they are one of the sources of contamination during the process.

Figure 9.1 shows a schematic diagram of a near-field scanning optical Raman microscope used for acquiring spectral data from samples. There has also been a great deal of research into the application of near-field scanning optical microscopy by many researchers, and these technologies include illumination mode, collection mode, and illumination–collection mode, among others. A near-field optical probe is formed by the formation of a fine electromagnetic field probe in the vicinity of a small aperture on the tip of the metal needle. While the size of the aperture controls the size of the near-field optical probe, the microscopic image is determined by the signal to noise ratio, which is dependent on the aperture size. There is a great deal of importance attached to both the probe size and the optical power throughput of the probe. The use of aperture-type near-field scanning optical microscopy has not yet been demonstrated to be able to achieve high spatial resolution (Mensi et al., 2010).

Several groups have reported the development of near-field optical Raman spectroscopy with improved illumination–correction modes and fine aperture pyramidal

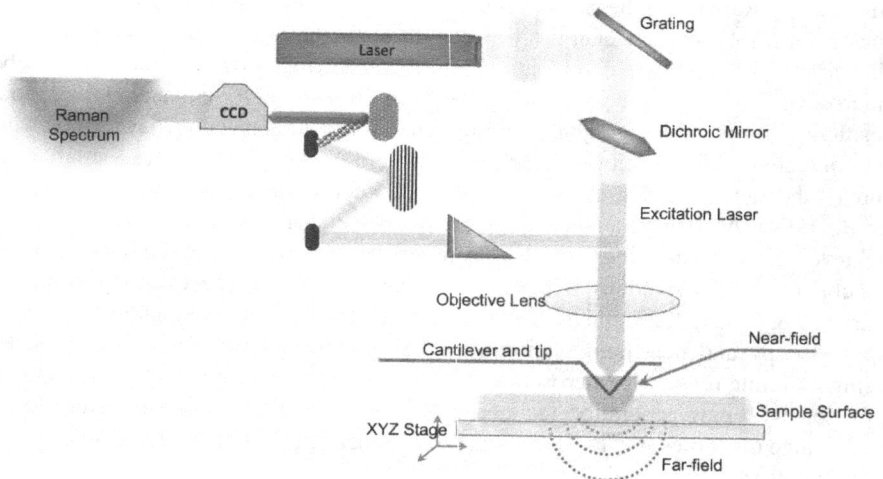

FIGURE 9.1 A schematic diagram of a near-field scanning optical Raman microscope used for acquiring spectral data and imaging from the sample surface.

probes. A Raman spectrometer and near-field scanning optical microscope have been coupled for detecting Raman spectral signals by integrating light and light scattering. There has been a realization that near-field Raman mapping is a powerful tool for studying the characteristics of materials, and devices. In order to characterize materials and devices on length scales of a few nanometers using visible light, near-field scanning optical microscopy can be combined with Raman scattering spectroscopy so as to provide a unique opportunity to identify distinct spectral features localized in the materials and devices.

9.2 WORKING PRINCIPLE

A near-field scanning optical microscopy Raman experiment can be performed on the basis of inelastic scattering in the region of interest. The typical experiment uses a laser to illuminate the sample with monochromatic light, which is applied to the surface of the sample. The use of laser sources as an excitation source is possible in the near-infrared, visible, and ultraviolet spectral ranges. When a monochromatic beam of light impinges on a sample, the photons of light collide with the vibrating molecules in the sample, resulting in scattered light. Based on the characteristics of the sample, either the scattered photon energy increases (or decreases) or does not change depending on the characteristics of the sample. In most cases, scattered radiation has the same frequency as the incident radiation and is therefore known as elastic scattering. The Rayleigh scattering process, an elastic scattering process, occurs when the frequency of scattered light equals the frequency of incident light, which results in the loss of no photon energy in the process (Paidi et al., 2020). The inelastic scattering of photons occurs when vibrational modes of atoms or molecules change, resulting either in a loss or gain of energy depending on the vibrational mode (Adamczyk et al., 2021).

The Raman scattering process is a phenomenon where there is only a very small fraction of scattered radiation that has a frequency different from the incident radiation, and this is what constitutes the Raman scattering process. There are two types of Raman scattering associated with the Raman effect: Stokes and anti-Stokes Raman scattering. According to Leon-Bejarano et al. (2017), Stokes Raman scattering is associated with energy loss, whereas anti-Stokes Raman scattering is associated with energy gain (León-Bejarano et al., 2017). A Raman spectrum is composed of Stokes scattering lines as well as anti-Stokes scattering lines. It has been found that most of the investigations have been focused on the Stokes region of the spectra on account of the considerable increase in availability of molecules in the ground vibrational state at ambient temperatures (Deinum et al., 1999; Paidi et al., 2020). There is no doubt that the wavelength of the incident radiation plays a crucial role in Raman scattering, and the Raman spectrum is produced when monochromatic light collides with a sample in an inelastic manner. In the next section, we will discuss the topic in a more detailed manner.

The inelastic scattering of light in Raman spectroscopy is analogous to the Compton scattering in X-ray experiments (Bergmann et al., 2002). It is a result of inelastic light scattering that the Raman effect is responsible for spectroscopy techniques that can be used in scientific applications to observe the vibrational and rotational modes in molecules by probing them. There is a difference between the wavelengths of the dispersed radiation emitted from the sample surface and the wavelengths of the illuminated monochromatic radiation emitted from the illuminated surface of the sample. Physicists Dr. K.S. Krishnan and Sir C.V. Raman, two famous Indian physicists, were the first to discover it in 1928 (Raman, 1928). When monochromatic light illuminates the sample, photons of light collide with molecules of the matter. Inelastic scattering occurs when only a few photons of incident light undergo inelastic scattering in all directions, depending on the characteristics of the sample type. In this process, photons either lose (Stokes shift) or gain (anti-Stokes shift) energy due to changes in the vibrational/rotational states of the constituent molecules of matter as a result of changes in their vibrational/rotational states. There is a phenomenon known as Rayleigh scattering which occurs when nearly all of the scattered light has the same frequency as the incident light, which is often the case in monochromatic light. There is a phenomenon known as elastic scattering in which photons of the same energy level scatter at the same rate as those of incident photons. There is, however, an inelastic scattering process when the energy level of the incident photon is different from that of the scattering photon (Rostron et al., 2016).

The Raman line spectrum is produced when inelastically scattered photons are detected from the surface of a molecule. There are different Raman bands associated with each component of a compound, and it is the intensity of these Raman bands that varies with the concentration of the compounds being analyzed in the particular sample. The compound has a spectrum for each of its components, as well as a Raman spectrum for each of the components of the compound. There is also a tiny energy difference between the incident and scattered photons, which is called the resonance Raman effect, which corresponds to Raman scattering. During scattering, there is a part of the radiation that has a wave number different from the part of the radiation that was incident. Based on the Raman scattering technique, which is also known as Raman scattering, the Raman method is based on the change in wavelength caused by the vibrational characteristics of the molecules responsible for scattering, as well as the component structure of the sample. Based on the variation in the wavelength of the scattered photons associated with the polarization of the molecule, it is possible to obtain structural and chemical information about the sample under evaluation by studying these variations.

It is a property of inelastic scattering that the change in molecular wavelength/ energy shift is represented as a "wave number" in terms of defining the intrinsic properties of the molecule such as its composition, coordination number, bonding of constituent atoms, or functional group of the molecule (de Juan et al., 2004). Figure 9.2 shows a description of the types of scattering that occur in a sample, and how scattering can be classified based on the quantification of energy that is being lost or gained by the sample. Based on the analysis of scattered photons, it is possible to classify the molecules into chemical, elemental, structural, and substructural categories as a result of the different types of bonds present in the sample. Due to the fact that Raman transitions do not require any molecular excitation in order to reach higher electronic states, they can be obtained in the visible and near-infrared regions with lower energy sources. It also minimizes the risk that the softer samples will be destroyed during the process (Esmonde-White et al., 2017).

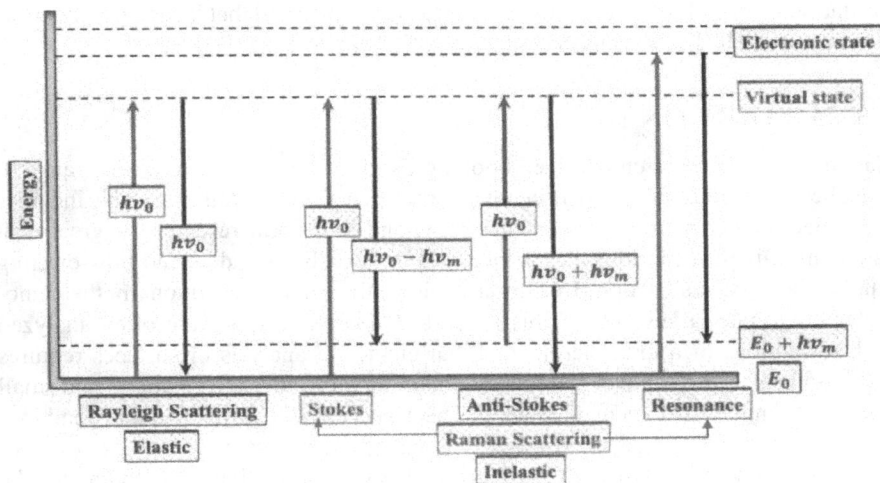

FIGURE 9.2 A description of the types of scattering that occur in a sample, and how scattering can be classified based on the quantification of energy that is being lost or gained by the sample.

The near-infrared region of the electromagnetic spectrum penetrates deeply into the subsurface of sample with low damage, but the Raman scattering intensity of the light decreases with an increase in wavelength. As a result, selecting an excitation source wavelength is a compromise between signal intensity and potential damage of softer samples. As one of the major challenges associated with near scanning optical Raman microscopy as a technique for imaging, this is one of the biggest challenges. Furthermore, the spectral characteristics of molecules that are generated by vibrational transitions are independent of the sources of laser excitation, which provides a more realistic picture of the spectral characteristics of molecules (Prats-Mateu & Gierlinger, 2017).

As a result of the considerably greater availability of molecules in the ground vibrational state at ambient temperature, most investigations have focused on the Stokes region of the spectra. Hence, Raman scattering is a spectroscopic technique used to gain information about the structural and chemical characteristics of molecules that are vital to the Raman spectroscopy technique (Schmitt & Popp, 2006). For both qualitative and quantitative analyses, Raman spectroscopy can be used to assess the concentrations of compounds in a given sample based on the band intensity in the Raman spectrum, and it can be used for both qualitative and quantitative analysis. There are a wide range of applications that have been discovered for Raman scattering since its discovery and have been discussed in the following section.

For example, Raman scattering can be used to perform chemical characterizations of a wide range of samples without using labels and without causing any damage to the samples in order to identify their chemical fingerprints. It has been reported that shorter wavelengths may increase the fluorescence background from the sample and decrease the Raman signal as a result. In previous research, it has been found that an excitation source of 532 nm works best for most samples and that it can directly probe a sample as well (Bøtter-Jensen et al., 2000; Castelletto et al., 2014).

9.3 APPLICATION

Nanometer-scale research is currently of a great deal of relevance to a wide range of branches of science and engineering that carry out research in this area. This includes microelectronics, supramolecular chemistry, and biological research. A great deal of attention is paid to molecular device design, usually to understand how existing "molecular systems" work. Detection tools with spatial resolution in the nanometer range play a key role in this process. Molecular species are often analyzed with resolutions of a few micrometers, but elemental analysis of surfaces requires resolutions of 1 μm. A novel method for obtaining molecular information from small samples as small as 50 nm in diameter has been successfully developed that combines scanning near-field optical microscopy with Raman spectroscopy.

It has been shown that the scattered intensity of an incident beam can be determined from the measurement of the polarization of the incident beam relative to the tip axis using apertureless scanning optical Raman microscopy (Nieman et al., 2001). The first study compared the performance of far-field and near-field Raman measurements using scanning optical Raman microscopy, carried out by Jahncke et al. (1995). $KTiOPO_4$ doped with Rb is imaged using Raman spectroscopy in conjunction with near-field scanning optical microscopy. The first Raman images and spectra obtained in the near field are presented (Jahncke et al., 1995). In a study, an apertureless near-field scanning optical microscope was used to demonstrate near-field Raman imaging of organic molecules that is based on a silver-layer-coated cantilever of an atomic force microscope as the tip. Raman scattering cross sections are enhanced because of the enhanced electric field that exists around the tip's apex due to the surface plasmon polariton excitations at the tip's apex. Through the use of near-field Raman images, it is possible to reveal the molecular vibrational distributions of Rhodamine6G and Crystal Violet molecules beyond the diffraction limit of a light source (Hayazawa et al., 2002).

The chemical composition of diamond films can be determined with a lateral resolution of less than 100 nm when using Raman spectroscopy through near-field microscopy, which can significantly improve the process of determining their purity and chemical composition. It is observed that a spectrum of a vapor-deposited diamond film has a sharp band at 1330 cm^{-1}, assigned to diamond, the broad feature at 1600 cm^{-1} assigned partially to glass and partially to graphitic carbon (Zenobi & Deckert, 2000).

9.4 SUMMARY

Near-field scanning optical Raman microscopy is a technique that provides not only high-resolution optical imaging, but also localized spectroscopic investigations of surfaces in the near field. With the help of this technique, it is possible to gain insights into molecular processes at the atomic level without any contact or invasiveness. In addition to its ability to detect sensitive molecules with spectroscopic fingerprints, the method also reveals the molecular structure, composition, and function in terms of size and structure. The technique is also capable of addressing some of the most compelling challenges in clinical translation and measurement in biomedical domains, such as medical diagnostics and disease monitoring that pose compelling challenges to clinical translation and measurement in medicine.

REFERENCES

Adamczyk, A., Matuszyk, E., Radwan, B., Rocchetti, S., Chlopicki, S., & Baranska, M. (2021). Toward Raman subcellular imaging of endothelial dysfunction. *Journal of Medicinal Chemistry*, *64*(8), 4396–4409. https://doi.org/10.1021/acs.jmedchem.1c00051

Atkin, J. M., Berweger, S., Jones, A. C., & Raschke, M. B. (2012). Nano-optical imaging and spectroscopy of order, phases, and domains in complex solids. *Advances in Physics*, *61*(6), 745–842.

Bergmann, U., Glatzel, P., & Cramer, S. P. (2002). Bulk-sensitive XAS characterization of light elements: From X-ray Raman scattering to X-ray Raman spectroscopy. *Microchemical Journal*, *71*(2–3), 221–230.

Bøtter-Jensen, L., Bulur, E., Duller, G. A. T., & Murray, A. S. (2000). Advances in luminescence instrument systems. *Radiation Measurements*, *32*(5–6), 523–528.

Castelletto, S., Johnson, B. C., Ivády, V., Stavrias, N., Umeda, T., Gali, A., & Ohshima, T. (2014). A silicon carbide room-temperature single-photon source. *Nature Materials*, *13*(2), 151–156.

de Juan, A., Tauler, R., Dyson, R., Marcolli, C., Rault, M., & Maeder, M. (2004). Spectroscopic imaging and chemometrics: A powerful combination for global and local sample analysis. *TrAC Trends in Analytical Chemistry*, *23*(1), 70–79.

Deinum, G., Rodriguez, D., Römer, T. J., Fitzmaurice, M., Kramer, J. R., & Feld, M. S. (1999). Histological classification of Raman spectra of human coronary artery atherosclerosis using principal component analysis. *Applied Spectroscopy*, *53*(8), 938–942. https://doi.org/10.1366/0003702991947829

Dong, B., Davis, J. L., Sun, C., & Zhang, H. F. (2018). Spectroscopic analysis beyond the diffraction limit. *The International Journal of Biochemistry & Cell Biology*, *101*, 113–117.

Esmonde-White, K. A., Cuellar, M., Uerpmann, C., Lenain, B., & Lewis, I. R. (2017). Raman spectroscopy as a process analytical technology for pharmaceutical manufacturing and bioprocessing. *Analytical and Bioanalytical Chemistry*, *409*(3), 637–649.

Hartschuh, A., Anderson, N., & Novotny, L. (2003). Near-field Raman spectroscopy using a sharp metal tip. *Journal of Microscopy*, *210*(3), 234–240. https://doi.org/https://doi.org/10.1046/j.1365-2818.2003.01137.x

Hayazawa, N., Inouye, Y., Sekkat, Z., & Kawata, S. (2002). Near-field Raman imaging of organic molecules by an apertureless metallic probe scanning optical microscope. *The Journal of Chemical Physics*, *117*(3), 1296–1301. https://doi.org/10.1063/1.1485731

Jahncke, C. L., Paesler, M. A., & Hallen, H. D. (1995). Raman imaging with near-field scanning optical microscopy. *Applied Physics Letters*, *67*(17), 2483–2485. https://doi.org/10.1063/1.114615

Krishna, R., & Colak, I. (2022). Advances in biomedical applications of Raman microscopy and data processing: A mini review. *Analytical Letters*, 1–42. https://doi.org/10.1080/00032719.2022.2094391

Krishna, R., Unsworth, T. J., & Edge, R. (2016). Raman spectroscopy and microscopy. In *Reference Module in Materials Science and Materials Engineering*. Saleem Hashmi ed., London: Elsevier. https://doi.org/https://doi.org/10.1016/B978-0-12-803581-8.03091-5

Lee, H., Yoo, H., Moon, G., Toh, K.-A., Mochizuki, K., Fujita, K., & Kim, D. (2021). Super-resolved Raman microscopy using random structured light illumination: Concept and feasibility. *The Journal of Chemical Physics*, *155*(14), 144202.

León-Bejarano, F., Ramírez-Elías, M., Mendez, M. O., Dorantes-Méndez, G., Rodríguez-Aranda, Ma. del C., & Alba, A. (2017). Denoising of Raman spectroscopy for biological samples based on empirical mode decomposition. *International Journal of Modern Physics C*, *28*(09), 1750116. https://doi.org/10.1142/S0129183117501169

Mensi, M., Mikhailov, G., Pyatkin, S., Adamcik, J., Sekatskii, S., & Dietler, G. (2010). Ultrasharp carbon whisker optical fiber probes for scanning near-field optical microscopy. *Nanophotonics III*, *7712*, 255–261.

Nieman, L. T., Krampert, G. M., & Martinez, R. E. (2001). An apertureless near-field scanning optical microscope and its application to surface-enhanced Raman spectroscopy and multiphoton fluorescence imaging. *Review of Scientific Instruments*, *72*(3), 1691–1699. https://doi.org/10.1063/1.1347975

Paidi, S. K., Pandey, R., & Barman, I. (2020). Emerging trends in biomedical imaging and disease diagnosis using Raman spectroscopy. In *Molecular and Laser Spectroscopy*, Advances and Applications: Volume 2, Elsevier, 623–652. ISBN 9780128188705, https://doi.org/10.1016/B978-0-12-818870-5.00018-6

Prats-Mateu, B., & Gierlinger, N. (2017). Tip in–light on: Advantages, challenges, and applications of combining AFM and Raman microscopy on biological samples. *Microscopy Research and Technique*, *80*(1), 30–40.

Raman, C. V. (1928). A new radiation. *Indian Journal of Physics*, *2*, 387–398.

Raman, C. V., & Krishnan, K. S. (1928). A new type of secondary radiation. *Nature*, *121*(3048), 501–502. https://doi.org/10.1038/121501c0

Rostron, P., Gaber, S., & Gaber, D. (2016). Raman spectroscopy, review. *Laser*, *21*, 24.

Schmitt, M., & Popp, J. (2006). Raman spectroscopy at the beginning of the twenty-first century. *Journal of Raman Spectroscopy: An International Journal for Original Work in All Aspects of Raman Spectroscopy, Including Higher Order Processes, and Also Brillouin and Rayleigh Scattering*, *37*(1–3), 20–28.

Trautman, J. K., Macklin, J. J., Brus, L. E., & Betzig, E. (1994). Near-field spectroscopy of single molecules at room temperature. *Nature*, *369*(6475), 40–42. https://doi.org/10.1038/369040a0

Zenobi, R., & Deckert, V. (2000). Scanning near-field optical microscopy and spectroscopy as a tool for chemical analysis. *Angewandte Chemie International Edition*, *39*(10), 1746–1756. https://doi.org/https://doi.org/10.1002/(SICI)1521-3773(20000515)39:10<1746::AID-ANIE1746>3.0.CO;2-Q

10 Optical Characterization Instruments

There is one primary objective of this chapter in the book, and that is to provide a brief introduction to the most commonly used optical characterization techniques and instruments. Optical characterization technique is a method of assessing the properties of a material through the use of photons of light in order to determine their quantitative properties. Many of these well-known characterization techniques, which are frequently used for the purpose of optical characterization, are based on the use of photons with energies that are part of the electromagnetic spectrum. They range in energy from 1.2 meV to 124 eV, and are commonly used to measure optical properties. A detailed description of each technique is provided, along with an explanation of the applicability, usefulness, and limitations of each technique. There will be a discussion about techniques that can be used in conjunction with or complement to each other and we will also discuss some of the hurdles that are commonly encountered when applying them in practical, real-world application examples as well as some suggestions as to how to avoid them as much as possible. The optical characterization techniques are usually nondestructive, fast, and easy to implement, with most requiring very simple preparation of the sample prior to the optical characterization process. As a result of interaction between the light photons and the material under investigation, the intensity, energy, phase, direction, or polarization of the light wave may change depending on the interaction. There are many of them that are able to be performed at room temperature and atmosphere, thus negating the need for complex vacuum chambers, enabling them to perform at any time.

It is important to point out that due to the remote nature of optical probes, it is possible to examine materials in situ during the processing, by focusing the beam of the probe into a small spot and scanning it, enabling different regions to be characterized and maps of the different properties to be produced during the process. The optical characterization methods include ellipsometry, infrared spectroscopy, interferometry, photoluminescence, optical microscopy (visible and infrared), photoreflectance, Raman scattering, reflectometry, and optical spectroscopy. Among them ellipsometry is one of the emerging optical characterization techniques.

DOI: 10.1201/9781003340546-11

10.1 CONSTRUCTION

Ellipsometry spectroscopy technique is an experimental method for measuring the amplitude ratio between two perpendicularly polarized beams of light reflected from a sample surface. This is done in order to determine the change in polarization of the light by measuring the amplitude ratio between these beams of light. The technique is based on sweeping the wavelength over a range of wavelengths (spectroscopic ellipsometry) from ultraviolet to infrared, which allows a more detailed analysis of complex structures, thickness, and optical properties of film materials (Tompkins & Irene, 2005). The importance of spectroscopic ellipsometry lies in the fact that it is capable of analyzing multiple layers and determining the optical constant dispersion (variation as a function of wavelength). By using the optical dispersion of a material, it is possible to deduce additional material properties, such as the degree of crystallinity present in the material. An ellipsometry probe consists of a polarized beam of light that is used to conduct the measurement. During the measurement, the sample of interest is illuminated by a light beam of known polarization; the beam of light is reflected by the sample and is then directed back into the instrument. The ellipsometry device measures the polarization state of reflected light beams as a consequence of what the sample did to them.

Ellipsometers are tools that can be used for measuring the thickness of films with an extremely high degree of accuracy and precision. Further, in the case of semi-transparent films, the transformation occurs when a beam of light is reflected from or transmitted through an interface or film (Riegler, 1993). It has become increasingly common for ellipsometry to be used as the primary method of determining the optical properties of materials, and it has been widely used in recent years to obtain intrinsic and structural properties in bulk and thin-film forms of homogeneous and inhomogeneous materials, along with surface and interface properties (Aspnes, 2013). In ellipsometry, model-based approaches are used as a means of determining surface roughness thicknesses, thin film roughness thicknesses, and interface roughness thicknesses, as well as optical properties of thin films ranging in thickness from a few nanometers to several tens of microns. The use of spectroscopic ellipsometry in a wide variety of applications can also be achieved either ex situ or in situ, in static or kinetic mode, depending on the application. The measurement of isotropic, non-absorbing layers is usually carried out by using a laser of a single wavelength (e.g., with the He-Ne laser which has a wavelength of 632.8 nm), which is usually used to measure these types of layers.

In order to obtain an accurate spectroscopic ellipsometer, it is essential to have a light source with a wide wavelength range. One of the most common sources of spectroscopic light is a lamp. As an example, arc lamps, such as Xenon and Deuterium lamps, function by energizing molecules within a plasma of ionized gas that emit light upon returning to their original state after being excited. Deuterium lamps produce light at ultraviolet wavelengths below about 400 nm. Xenon lamps can produce light with wavelengths ranging from 185 nm to 2000 nm. The wavelengths of light emitted by halogen lamps extend into the near infrared, exceeding 350 nm. Mid-infrared wavelengths are covered by silicon carbide globars. It is important to note that the output of each lamp can vary widely within its range of useable wavelengths.

A crucial aspect of an ellipsometry experiment is the availability of light at all of the wavelengths that are measured, and the absence of any fluctuations in intensity over a period of time separate from that of the measurement period.

There is a detector embedded in every ellipsometer, which is responsible for converting the electromagnetic field into a measurable property. Generally, this is done using a photodiode detector or an array of photodiodes. The photodetection process involves converting the electromagnetic waves into a voltage or current that is proportional to the amount of light detected. Depending on the wavelength range, the type of photodetector and detector material used will vary. There is a direct correlation between the operating wavelengths of semiconductor detectors and the bandgaps of the materials. In order to generate an electronic transition within a material, the bandgap energy must have a minimum photon energy that is sufficient to excite an electronic transition within the material, and it is inversely proportional to the wavelength of light.

The photomultiplier tube (PMT) is also one of the types of detectors used in spectroscopic ellipsometers. The device is based on a photosensitive material, called a photocathode, that will release electrons upon being struck by a photon (light). In very low light conditions, PMT detectors are an excellent choice. Photocathode materials have a tremendous effect on the wavelength range that can be achieved with a PMT.

10.2   WORKING PRINCIPLE

Using an ellipsometer, the entities that are measured are the reciprocally perpendicular components of the probing beam (called the p-waves and the s-waves) that are perpendicular to each other. It is the ratio between the amplitudes of the mutually perpendicular components that gives rise to the quantity called psi (ψ), and the phase shift between the two components that gives rise to the quantity called delta (Δ). There are a variety of wavelengths of light that are measured for ψ and Δ, which is why it is referred to as spectroscopic technique. The effect of ψ and Δ is such that linearly polarized light will be altered to elliptically polarized light upon reflection from the surface of a sample, as shown in Figure 10.1 (Tompkins & Hilfiker, 2015).

An ellipsometry function such as the thickness of the film (or substrate) and the optical properties of the film (or substrate) can be calculated via software models by converting the ellipsometry quantities ψ and Δ into the desired ultimate quantities. It is pertinent to note that the measured response is dependent on the optical properties and thickness of the individual materials. As a result, ellipsometry is primarily used to determine the thickness of films as well as their optical constants. However, it is also effective for characterizing the composition, crystallinity, roughness, doping concentration, and other material properties that are associated with a change in optical response for a particular material. The prevalent use of ellipsometry in a wide range of areas can be explained by the increased reliance on thin films in many areas, as well as the flexibility of ellipsometry to measure a wide range of materials: dielectrics, semiconductors, metals, superconductors, organics, biological coatings, and composite materials.

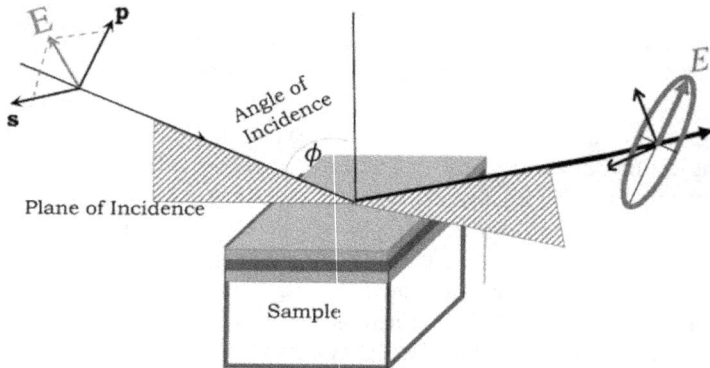

FIGURE 10.1 With both p- and s-waves components of linearly polarized light incident, the ellipsometry measurement was shown to be possible with linearly polarized light. In the presence of a sample, the incident polarizations are changed in amplitude and phase for p- and s-polarizations, resulting in the production of elliptically polarized light.

The simplest form of light can be thought of as a wave of electromagnetic energy traveling through space. This waveform can be used to explore the behavior of the wave's electric fields in space and time. This behavior is also known as polarization. An electric field associated with a wave is always orthogonal to the direction of propagation of the wave. This means that a wave traveling in the z-direction can be described by the x- and y-components that make up the wave.

There is no polarization of light when the orientation and phase of the light are both completely random. In order to understand ellipsometry, it is necessary to understand the kind of electric field that falls into a specific path and traces out a distinct shape at any point along the way. As a result, polarized light is produced. In the event that two orthogonal light waves are in phase, the resulting light will have a linear polarization. Depending on the relative magnitudes of the amplitudes, the orientation that results is determined. In the case where the orthogonal waves are 90° out of phase and have equal amplitudes, the resultant light is circularly polarized. Among the most common forms of polarization, "elliptical" is the most commonly used, which combines orthogonal waves of different amplitudes and phases in a single wave. As a result of this, the name "ellipsometry" was given. There are therefore optical elements that must be employed to manipulate light polarization – both in order to form a known polarization before the sample and in order to detect the polarization after the sample. It is imperative to understand the key principles of polarizing optics, which are used by spectroscopic ellipsometers. The ellipsometer may contain additional optical elements, such as lenses, mirrors, and pinholes, in addition to the optical elements mentioned above.

An ellipsometer consists of two main parts based on the sample that is being measured. Polarization state generators are a group of optical elements that are placed before the sample and have the general task of creating a polarization state that can be used to interact with the sample, hence they are collectively referred to as polarization

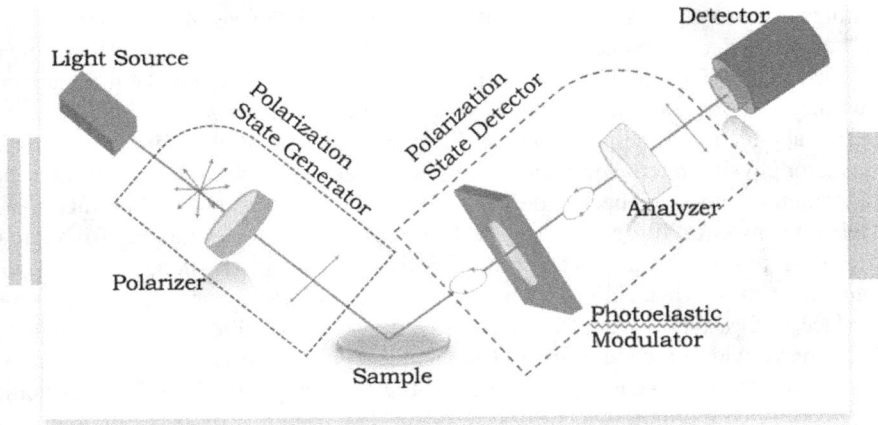

FIGURE 10.2 As light passes through the polarization state generator, it is polarized according to the angle of incidence at which it reflects from the sample surface. The light is reflected by the polarization state detector, which determines the polarization change caused by interaction with the sample.

state sources. As a result of an interaction between the polarized light and the sample (reflection or transmission), the original polarization of the light is changed. It is necessary to determine the cause of the change in polarization in order to identify the sample properties that caused the change. Therefore, the optical components are used to detect the new "unknown" polarization state of the sample. Polarization state detectors are generally recognized as a collective term for these detectors. Figure 10.2 illustrates that ellipsometry has a number of components.

There is evidence that ellipsometry can produce information about layers that are thinner than the wavelength of the probing light itself, even down to the level of a single atomic layer. The nondestructive, contactless characteristics of ellipsometry as an optical technique make it desirable for the inspection of soft layers on biomaterials since the technique is not destructive. The ellipsometry method is also less affected by intensity instability of the light source or atmospheric absorption, since it measures the intensity ratio instead of pure intensity, and there is no requirement for reference measurements, so relatively low accuracies can be achieved when using this method.

10.3 APPLICATION

The optical characterization technique is used for a number of purposes, such as determining the thickness or optical constant of thin films, monitoring chemical reactions and film growth, studying the plasmonic effects in meta-materials, and analyzing the structure of bio- and nanomaterials (Arwin, 2011; Balevicius et al., 2014). There is no doubt that the range of applications of this technology has expanded significantly in recent years. This includes basic research in physics to semiconductor and data storage solutions, flat panel displays, communication, biosensors, and optical coatings. Ellipsometry is sensitive to thin films over a thickness of a few nanometers and can be used as a method of in situ observation of passive films in electrochemical systems. In combination with electrochemistry, it allows us to study the mechanisms and kinetics of the formation of oxide films on metal surfaces.

This application of ellipsometry can be seen in a wide variety of fields like semiconductor physics, microelectronics, biomaterials, etc. In the field of biomaterials, the ellipsometry method has been widely used to determine the thickness of biofunctional layers on substrates. It is commonly used to measure thicknesses ranging from a few Angstroms to several micrometers for layers that are optically homogeneous and isotropic, as well as when a significant discontinuity in refractive indexes exists at the interface of the layers (Wang & Chu, 2013). It is even possible to perform imaging ellipsometry with a monochromatic laser as the light source and a CCD camera as the detector, thereby extending the concept of ellipsometry into a new field (Szmodis et al., 2008). In the field of materials research, ellipsometry spectroscopy is one of the most widely used methods and has become a routine tool for characterization. The method is particularly sensitive to impurities, and its instrumentation is relatively simple and inexpensive to use. The Raman scattering technique is very useful for characterizing lattices and their stresses, impurities, and free carriers. It is generally agreed that photoluminescence and Raman scattering function as surface probes, since they are able to penetrate only tens to hundreds of nanometers into semiconductor materials. Recently, advances in the use of optical methods are leading to advances in the characterization of stress dead zones in fatigued specimens that are under uniaxial tensile loading. Using high-resolution optical techniques, such as digital image correlation and electronic speckle pattern interferometry, the authors were able to produce a high-resolution image (Farahani et al., 2022).

10.4   SUMMARY

The optical characterization techniques stand out from the rest of the characterization methods because they require less preparation of the samples, for example, the time-consuming process of creating electrical contacts is eliminated in comparison to other characterization methods. The sample is therefore generally not altered, nor does the measurement itself cause any damage unless the laser beam used for probing is too intense, which can be avoided in most situations. This technique can be used to examine different parts of the structure with spatial resolutions determined by the wavelength of light being used. Optical beams are easy to manipulate, allowing them to examine different parts of structures. It is possible to probe the finest details of a semiconductor microstructure or device using visible or near-infrared light. In other words, optical measurements can produce two-dimensional maps of sample properties, such as impurity distribution or layer thickness. In addition, it is also possible to differentiate properties along the third dimension by propagating light perpendicular to the surface of the sample. Light penetrates from nanometers to micrometers deep depending on its wavelength and sample properties. In light of the above reasons, optical characterization techniques are regarded by many as a valuable addition to the array of analytical tools available for materials analysis. One optical method of ellipsometry is discussed in this chapter, in which a known polarized light beam illuminates a sample surface, and the reflected light beam's polarization state is measured. It is widely recognized that these techniques are being used in research, material development, and the semiconductor and device industries. It is a spectroscopic technique that measures intensity versus wavelength, thereby allowing quantitative analysis and measurement of properties. As a general rule, photoluminescence and Raman scattering are typically excited in the near ultraviolet, visible, and near-infrared range between the wavelengths of 0.4 and 1μm. Therefore, among the other optical techniques used to measure thin films and bulk materials, ellipsometry is one of the most commonly used. By using the polarization changes caused by reflection or transmission from a material structure, it is possible to deduce properties of that material, such as the thickness and physical constants, based on the polarization changes. There is an increasing demand for thin film and material characterization techniques as the demand for thin film and optical characterization continues to rise.

REFERENCES

Arwin, H. (2011). Application of ellipsometry techniques to biological materials. *Thin Solid Films*, *519*(9), 2589–2592. https://doi.org/https://doi.org/10.1016/j.tsf.2010.11.082

Aspnes, D. E. (2013). Spectroscopic ellipsometry—A perspective. *Journal of Vacuum Science & Technology A*, *31*(5), 058502. https://doi.org/10.1116/1.4809747

Balevicius, Z., Baleviciute, I., Tumenas, S., Tamosaitis, L., Stirke, A., Makaraviciute, A., Ramanaviciene, A., & Ramanavicius, A. (2014). In situ study of ligand–receptor interaction by total internal reflection ellipsometry. *Thin Solid Films*, *571*, 744–748. https://doi.org/https://doi.org/10.1016/j.tsf.2013.10.090

Farahani, B. V., Direito, F., Sousa, P. J., Melo, F. Q., Tavares, P. J., Infante, V., & Moreira, P. M. G. P. (2022). Advancement on optical methods in stress dead-zone characterisation and SIF evaluation. *Engineering Failure Analysis*, *140*, 106493. https://doi.org/https://doi.org/10.1016/j.engfailanal.2022.106493

Riegler, H. (1993). A user's guide to ellipsometry. By Harland G. Tompkins, Academic Press, New York 1993, 260 pp. hardback, ISBN 0-12-603050-0. *Advanced Materials*, *5*(10), 778. https://doi.org/https://doi.org/10.1002/adma.19930051034

Szmodis, A. W., Blanchette, C. D., Levchenko, A. A., Navrotsky, A., Longo, M. L., Orme, C. A., & Parikh, A. N. (2008). Direct visualization of phase transition dynamics in binary supported phospholipid bilayers using imaging ellipsometry. *Soft Matter*, *4*(6), 1161–1164. https://doi.org/10.1039/B801390J

Tompkins, H. G., & Hilfiker, J. N. (2015). *Spectroscopic Ellipsometry: Practical Application to Thin Film Characterization*. Momentum Press.

Tompkins, H., & Irene, E. A. (2005). *Handbook of Ellipsometry*. William Andrew.

Wang, H., & Chu, P. K. (2013). Chapter 4–Surface characterization of biomaterials. In A. Bandyopadhyay & S. Bose (Eds.), *Characterization of biomaterials* (pp. 105–174). Academic Press. https://doi.org/https://doi.org/10.1016/B978-0-12-415800-9.00004-8

11 Synchrotron Techniques

11.1 PROTEIN CRYSTALLOGRAPHY

Protein crystallography is a very high-resolution form of microscopy, and this form of microscopy allows the scientist to see molecules at atomic scales (Stubbs, 2007). It provides us with the ability to see beyond the capability of even the most powerful light microscopes. All biological processes are based upon proteins. Being able to see the fine details of their structures can be extremely insightful as they are the basis of every biological process (Hogg & Hilgenfeld, 2007). The study of protein crystallography, for example, might enable us to get a better idea of why a change in the structure of a protein has such a dramatic impact. A second application of this technique would be to use it to provide us with information about the active site of an enzyme. This would enable us to make molecules that could target that area of the enzyme and provide a basis for developing new medicines (Joosten et al., 2007). In the past few decades, X-ray crystallography has been shown to be an incredibly successful technique in analyzing protein structure.

Initially, there were several difficulties related to the use of synchrotron radiation as a research tool when it comes to the collection of data from protein crystals; these include the vast amounts of diffraction data, the inherent weakness of the intensity of each reflection, the possibility of radiation damage on the sample, and the phase problem. With the advancement of technology, the synchrotron radiation has made a special contribution to technological progress (Moreno, 2017).

The purpose of protein crystallography is to construct atomic-resolution models of proteins using the techniques of X-ray diffraction. Post-genomics research is attracting a lot of attention and funding for protein molecules, which are complex three-dimensional structures. It is clear from the ethnographic analysis of protein crystallography that to become a specialist crystallographer, and therefore able to make sense of such intricate objects, researchers need to draw on their bodies as a resource to learn about, work with, as well as communicate precise molecular configurations as they become expert crystallographers. As a consequence of the heavy reliance on computer graphics technology used in contemporary crystallographic modeling, it is relatively difficult to build a model without actively handling it and manipulating it on screen throughout the many-hour long process of building the model. As part of a recent paper, the authors examine the body of works of crystallographic modeling

DOI: 10.1201/9781003340546-12

(Myers, 2008). In particular, they examine the corporeal practices through which modelers learn the intricate structures of protein molecules. They do this by drawing on ethnographic observations of contemporary protein crystallographers and historical accounts of early molecular modeling techniques (Law, 1973).

An ethnographic study suggests that, in addition to the information gathered from ethnographic studies, crystallographers sculpt embodied models alongside what they are building digitally on the screen. This is part of the process of building and manipulating protein models. It is not only a means of representing proteins at the computer interface that crystallographic modeling at the computer interface serves to produce; it is also a means of training novice crystallographers' bodies and imaginations with regard to crystallographic patterns. It has been suggested that the molecular embodiments used by protein crystallographers present a challenging way to investigate the visual cultures and knowledge practices in the computer-mediated life sciences by presenting a unique set of questions.

During the past few years, it has become easier and easier to obtain single-crystal diffraction facilities with synchrotron radiation for use in protein crystallography, with access to several suitable beamlines around the world, and this has led to considerable developments in the field. Figure 11.1 illustrates the schematic diagram of synchrotron radiation source.

A synchrotron is a source of radiation that has several properties due to electron motion caused by relativistic electrons in simple circular orbits in bending magnets, as well as by motions larger and more complex produced by electrons in wiggler or undulator magnets. There are a number of synchrotron beamlines that can produce intense X-rays, making them a powerful tool for macromolecular crystallographers in their pursuit of understanding biological phenomena at an atomic resolution. In protein crystallography, newer high brightness synchrotron radiation sources have had a significant impact. There are a number of instruments that are required to

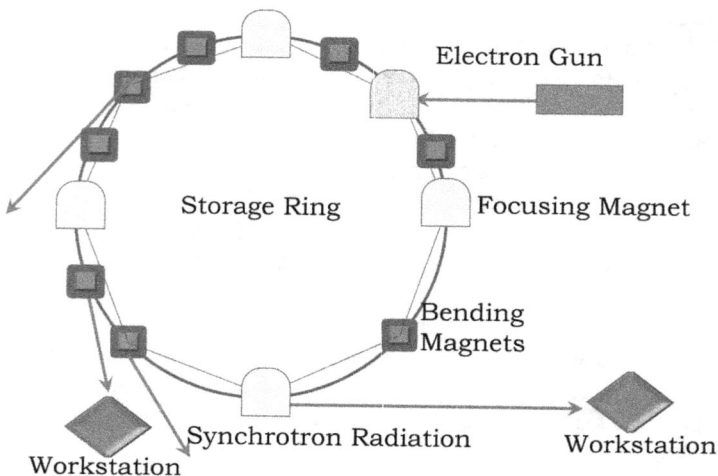

FIGURE 11.1 Schematic representation of the synchrotron radiation source.

monochromatize and focus the radiation onto typically small protein crystal samples in order to detect the diffraction pattern as a result. Over the last 30 years, the "golden age" of structural biology has been defined by the insights gained into the mechanisms' underlying biological processes that have played an essential role and shaped the field of biomedical sciences (Grabowski et al., 2021).

As a result, it is possible to determine a complete crystal structure for a wider range of materials, particularly for microcrystalline samples, than could previously have been possible. Special features of synchrotron radiation such as X-ray wavelength tunability can be exploited. For chemists interested in structural characterization, synchrotron radiation plays an important role, as demonstrated by these studies. Furthermore, it presents examples of results from various areas of concern to chemists, including microporous materials, pharmaceuticals, and supramolecular chemistry (Clegg, 2000).

In spite of the impressive advances in synchrotron technology over the past 30 years, the median resolution of macromolecular structures that have been determined using synchrotrons remains consistent over time at just about 2 Å. The present study suggests that synchrotrons of the future will increasingly move toward becoming a life science center model, in which X-ray crystallography, Cryogenic Electron Microscopy (cryo-EM), and other experimental and computational resources are incorporated within a versatile facility, capable of integrating a wide range of science. In light of the recent response of crystallographers, it is reasonable to conclude that the use of synchrotron beamlines for X-ray protein crystallography will continue to play an essential role in structural biology as well as drug discovery for a long time to come (Grabowski et al., 2021; Rotella et al., 2011).

11.2 X-RAY SCATTERING

As a family of non-destructive analytical techniques, X-ray scattering techniques can provide information about the crystal structure, chemical composition, and physical properties of materials and thin films without causing any damage to the material. As a basis for these techniques, the scattered intensity of an X-ray beam hitting a sample is observed as a function of angles of incident and scattered light, polarizations, wavelengths, and energy levels as a function of these variables. The incident X-rays from an X-ray source can be deflected and scattered by a sample when such a source is illuminated by X-rays, creating complex patterns on the sample. There are several types of X-ray scattering techniques available to reveal structural, elemental, and atomic information about a sample, including the X-ray scattering technique, which can be characterized by the intensity of the scatter (incident vs. scattered X-rays), polarization, wavelength, or energy changes, and is also known as X-ray scattering (Xiao & Lu, 2019).

As a research tool, X-ray scattering can be used to study a wide range of samples, from simple crystallography and novel material development to biological molecules and polymers that are complex. In addition to X-ray microscopy and X-ray spectroscopy, these techniques can also be used to probe samples, as they are nondestructive and can be used in conjunction with other methods. By using X-ray scattering, samples can be analyzed if they are capable of dispersing X-rays when they are exposed to

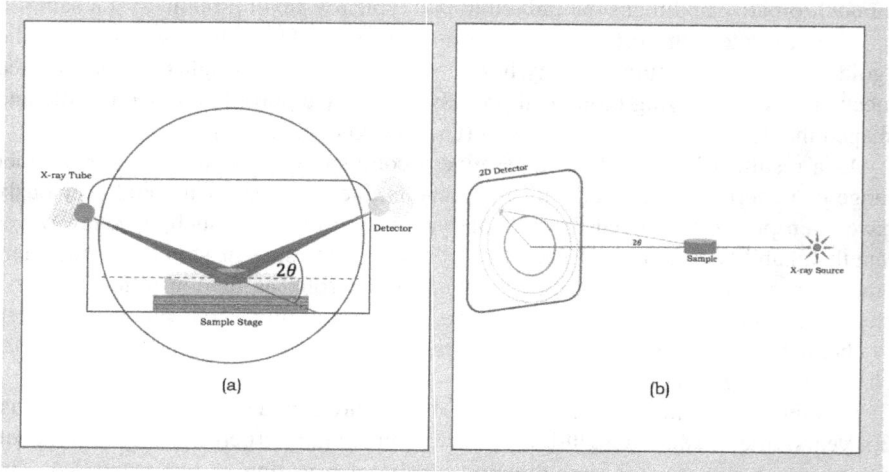

FIGURE 11.2 (a) Schematic X-ray scattering from the sample and (b) two-dimensional X-ray scattering technique.

them. It is possible to measure a variety of properties with X-ray scattering, including the dimension and shape of a sample, dispersity (the distribution of the size and shape of molecules within a sample), porosity, morphology, and orientation of the sample, among others. Figure 11.2 illustrates (a) schematic X-ray scattering from the sample and (b) two-dimensional X-ray scattering technique.

It is expected that X-rays scattered by electrons within a sample will interfere with each other when they are in an area of dense electrons; therefore, detectors usually analyze and characterize scattering within the sample that has been subject to constructive interference so that the electron location and other structure of the sample can be identified. In comparison with more structured samples such as fibbers and crystals, powders and dispersions display more random scattering patterns, while crystals exhibit highly anisotropic, separate scattering patterns that are similar to diffraction patterns (Liu et al., 2022).

11.2.1 SCATTERING TECHNIQUES BASED ON X-RAYS

As a general rule, X-ray scattering techniques can be classified into elastic and inelastic scattering techniques. It should be noted that elastic scattering produces scattered X-rays that have the same energy/wavelength as the incident X-rays, whereas inelastic scattering produces scattered X-rays that have an energy/wavelength different from the incident X-rays. The X-ray diffraction/crystallography technique is one of the most well-known elastic scattering techniques, including small-angle X-ray scattering (SAXS) and wide-angle X-ray scattering (WAXS). There are two subcategories of X-ray scattering that use crystalline samples in order to determine detailed information on the atomic level which require extremely high energy hard X-rays (with very small wavelengths) to be used. Both techniques are subsets of X-ray scattering. An example of an inelastic technique is Raman scattering and resonant inelastic X-ray scattering, among others.

The X-ray scattering technique has been used in a variety of fields of study to characterize the structure of materials. The unique aspect of this method is that it can be used for examining materials under realistic sample environments in real time. Through the use of complementary SAXS and WAXS, researchers can gain a deeper understanding of the physical properties of materials at nanometer and angstrom lengths (Li et al., 2016). It is possible to gain different sample information at different resolutions by detecting X-rays scattered at different angles, and detecting each angle separately. The SAXS technique is a nondestructive method and it is relatively simple and fast to carry out the experiments involved in the scattering process. The scattering angle is the angle between the original direction of the photons and the deviation caused by the deviation from the original direction. Obviously, the greater the scattering object, the smaller will be the scattering angle.

For this reason, this technique is called small angle scattering, and is in contrast to wide angle scattering, which is used to study atomic distances since the scattering angle is larger. It is an analytical technique which measures the intensity of the X-rays scattered by a sample as a function of how much angle of scattering it has. The measurements are usually taken at angles ranging from $0.1°$ to $5°$ in a range of very small angles. When a material contains structural features that measure on the nanometer scale, usually between 1 and 10 nm, a SAXS signal is observed. Loktesva et al. (2021) conducted real-time X-ray scattering to discover rich phase behavior in PbS nanocrystal superlattices during in situ assembly. Czajka and Armes (2021) studied time-resolved SAXS during aqueous emulsion polymerization to get useful insights into the mechanism of emulsion polymerization. Raschpichler et al. (2020) reported WAXS on free silica particles of different porosities prepared in a beam. A characteristic difference between the WAXS patterns of particle samples of different surface properties and porosity was reported (Raschpichler et al., 2020).

11.3 X-RAY ABSORPTION SPECTROSCOPY (XAS)

XAS is one of the most widely used techniques for determining the local geometrical or electronic structure of a substance. In most cases, the experiment is carried out at synchrotron radiation facilities, which can provide intense and tunable X-ray beams for the experiment to be conducted at a synchrotron source. It is possible to analyze gaseous, liquid, or solid samples depending on the type of sample being analyzed. In the study of the morphological details of chemical substances, XAS has emerged as one of the most important tools. In terms of electromagnetic radiation, X-rays are short-wavelength electromagnetic waves. There are various ways in which they can be used to analyze structural arrangements. It is because of their high energy that they are able to penetrate deep down to the core electrons and excite them to a higher level. It has been demonstrated over the past 30 years that X-ray absorption spectroscopy has made major contributions to a wide range of material research in a wide variety of fields (Sauer et al., 2008). The review of biochemical research including photosynthesis and oxygen evolution has been conducted a number of times (Sauer et al., 2005).

Thus, X-ray absorption is a synchrotron-based characterization technique that can be divided into near-edge spectroscopy (XANES) and extended X-ray absorption fine structure (EXAFS). In X-ray spectroscopy, electrons orbiting in the lower shells of the atom are excited based on the principle of excitability that occurs at the core of the electron. The electron absorbs X-rays and thus becomes excited and jumps to a whole new level as a result of becoming excited. In the X-ray region, the wavelengths range from 1 to 100 nm. As a result of the interaction between X-rays and electrons, electrons are excited to higher levels of energy. With the help of an X-ray absorption spectrum, one is able to distinguish between various elements based on their characteristic value of energy absorbed by the electron (de Groot, 2001).

Here is a brief description of the main components of an X-ray absorption spectrophotometer:

X-ray generators – These generators create X-rays by striking metal plates to knock out the electrons in the core. A wavelength range between 0.1 and 100 nm is considered to be the wavelength range of short wavelengths and high-energy waves, and they are called X-rays.

Monochromator – A monochromator is a device that divides the wavelengths produced by the X-ray source into a series of wavelengths that can be filtered by a monochromator. This ensures that only one wavelength approaches the target at a time, making sure that only one wavelength approaches the target as needed. Collimator – As a result of its alignment and narrowing properties, the beam can be aligned and narrowed down to 0.1 mm in diameter. In order to focus a crystal more precisely, a collimator is the most common reason to use one.

X-ray detector – An X-ray is a radiation that has the ability to cause analytes to absorb part of its energy and reflect the rest toward detectors. In order to detect the X-rays, the following methods can be used:

(a) An electron-hole diode is a semiconductor device that contains electrons that are paired with holes. They change their ratio when they come into contact with light. A pulse can be created by converting this change into a voltage.

(b) In a photomultiplier tube, in order to make the absorption spectrum of the light signal easily readable, this light signal is converted into an electric signal and is multiplied thousands of times before being electronically enhanced for easy reading on these detectors (Yano & Yachandra, 2009).

Figure 11.3 illustrates a schematic diagram of XAS.

The advantages of XAS are that it is extremely useful for characterizing the chemical nature and environment of atoms in molecules because of the element specificity of the process, and synchrotron sources provide X-ray energies that are appropriate for a wide range of elements on the periodic table – in particular, those found in metalloenzymes that are redox-active. In most cases, the X-ray energy that is used as a basis for the selection of the probe determines the type of element that will be studied (de Groot, 2001).

There is a broad definition that states that XAS may be defined as the study or use of X-ray absorption in relation to X-ray energy. It is an edge when, at a specific

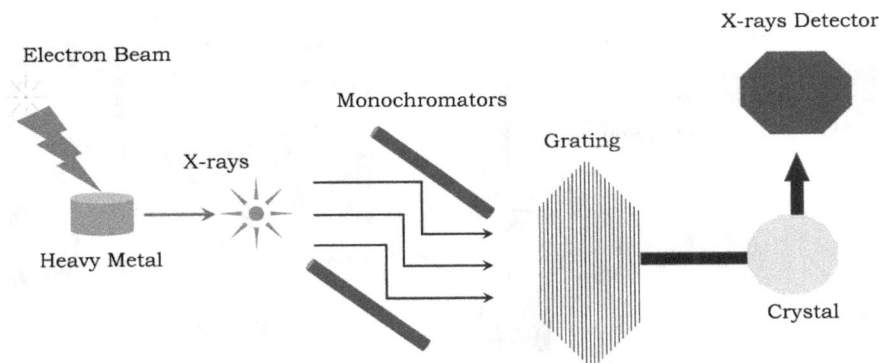

FIGURE 11.3   Schematic diagram of X-ray absorption spectroscopy.

value, the probability of absorption increases suddenly, which we call an absorption threshold. The energy of each edge depends primarily on the atomic number of the absorbing atom, as well as on how fast it absorbs the light. EXAFS is a technique that uses X-ray absorption spectra to investigate a wide range of energy levels, from the rising edge of the spectrum to some value, sometimes covering more than 1000 eV. It is observed that the absorption rate gradually decreases as the energy increases, but with oscillatory structures superimposed on top of the decrease in the absorption rate (Wende, 2004).

It has been observed that XAS has had a positive impact on a variety of scientific disciplines, attributed to the following aspects: (a) the XAS spectra can be quantitatively analyzed through the application of a solid theory; (b) there has been an increase in the availability of beamlines at synchrotron radiation facilities that allow high-quality XAS spectra to be collected; and in addition, (c) there is a need to develop reliable codes that can be applied to the analysis of data in a safe, reproducible, and controlled manner (Penner-Hahn, 2003).

In addition to being able to detect elements based on their core levels, the graph of the elements is well separated as well as having a high level of resolution. In this respect, it is applicable to both qualitative and quantitative analyses, which makes it a versatile tool. In addition to determining the structure of solids in metal alloys, this can be used to determine their orientations. Furthermore, it is also able to provide us with information as to what happened during the time when an electron jumped and came back. In explorations, XAS is a more sensitive technique when studying the structure and is widely employed in the study of the structure (Natoli et al., 2003).

11.4   PHOTOELECTRON SPECTROSCOPY

Depending on the type of solid, gas, or liquid, photoelectron spectroscopy is used to measure the energy of photoelectrons emitted from the solid/gas/liquid based on the photoelectric effect. Photoelectron spectroscopy can be divided into two main groups – ultraviolet photoelectron spectroscopy (UPS) and X-ray photoelectron spectroscopy (XPS) – based on the source of the ionization energy. It is a noble gas

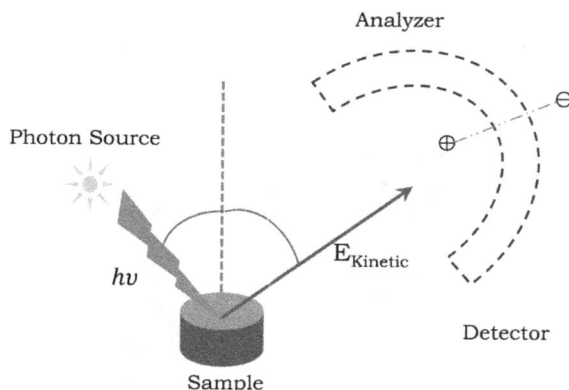

FIGURE 11.4 Principle of a photoemission spectroscopy. Monochromatic photons with energy *hv* are produced by a light source.

discharge lamp, usually a He discharge lamp, which produces the radiation for UPSs. A high energy X-ray beam (1000–1500 eV) is used as the source of energy for XPS, also known as electron spectroscopy for chemical analysis. It is important to note that in UPS, eject electrons come from the valence electrons, while in XPS, eject electrons come from the core electrons. In recent years, photoelectron spectroscopy (PES) has provided fundamental insight into the physical properties and chemical composition of surfaces and interfaces (Holmes, 2017). Figure 11.4 shows the principle of photoemission spectroscopy (Reinert & Hüfner, 2005a).

In terms of determining the electronic structure of organic and inorganic molecules, PES provides some of the most detailed quantitative information about the ionization energy, also known as electron binding energy. Essentially, ionization is the process by which a neutral molecule transitions from its ground state to its ion state. The energy of ionization can be divided into two types: the energy of adiabatic ionization and the energy of vertical ionization. The adiabatic ionization energy of a molecule is defined as the amount of energy required to eject an electron from the neutral molecule as a minimum. As an additional definition, this term can be considered as a measure of the difference between the vibrational ground state of a neutral molecule and the ground state of a positive ion. In relation to the second type of ionization energy – vertical ionization energy – it is accounted for any additional transitions between the ground and excited vibrational state of the neutral molecule. It is most likely that the transition will take place at the vertical ionization energy.

The excited species will eject electrons from the molecules as soon as they are excited by a photon of high energy in the ultraviolet region of the spectrum, which is enough energy to ionize the molecules. In PES, the kinetic energies of the electrons that are ejected are analyzed. An ejected electron's energy distribution is a reflection of the energy levels of the excited (ionized) molecule at a given excitation energy level for a given excitation energy. Photoionization transitions can be explained by the Franck–Condon principle, which explains the relative intensities of the vibrational bands. It has been shown by Koopman's theorem that the negative eigenvalue of an occupied orbital derived from a Hartree–Fock calculation is equal to the vertical ionization energy of the ion state that is formed after the photoionization of a molecule (Reinert & Hüfner, 2005). Ionization energies are shown to be closely related to molecular orbital energies, as well as ionization energies being directly related to the energies of molecular orbitals. However, there are some limitations to the theorem (Green, 2007). The PES method has thus been established as one of the most important methods for studying the electronic structure of molecules, solids and surfaces in the context of chemistry. A major benefit of PES is that it has been shown to have implications in numerous fields, such as surface chemistry or materials science, and has made a significant contribution to our understanding of the fundamental principles of solid state physics (Hüfner, 2003).

Photoelectronic excitation with synchrotron radiation has become more and more important in recent years because it allows measurements that cannot be performed with conventional UV or X-ray sources in the laboratory, due to the fact that synchrotron radiation can excite photoelectrons. One of the main differences between these sources and laboratory sources is that the photon energy can be selected by the use of a monochromator, which is capable of separating out a continuous spectrum of photons across a wide range of energy levels. The synchrotron light offers a number of other advantages, such as high intensity and brightness, variable polarization, small photon spots, or the possibility of time-resolved measurements on a very short timescale such as nanoseconds or even less. For many applications, however, the use of the relatively simple and cheap laboratory sources is still preferable than the use of more complicated, expensive sources (Reinert & Hüfner, 2005).

This method is widely used in particular for analyzing solids and surfaces, and it has made a significant contribution to the understanding of the electronic structure of condensed matter. PES has to an important extent contributed to our current understanding of electronic band structures and Fermi surfaces, which are largely derived from experimental data obtained by PES over the past few decades. It is common knowledge that photoelectron microscopy is now one of the most widely used analytical methods in materials science and chemistry, and it is gaining more and more importance in fields such as nanotechnology and biology as well.

11.5 INFRARED SPECTROSCOPY

The use of infrared (IR) spectroscopy is one of the most widely applied analytical techniques and has the benefit of being capable of analyzing virtually any sample in almost any state. With a judicious choice of sample sampling techniques, liquids, solutions, pastes, powders, films, fibers, gases, and surfaces can all be examined based on their composition and properties. Inorganic and organic chemists mostly utilize this technique as one of the most common and widely used spectroscopic techniques because of its usefulness in determining the structure of compounds as well as identifying them. A spectrophotometer is an instrument that can be used for generating a spectrum of IR light by using an instrument called an infrared spectrometer (Ng & Simmons, 1999).

A common method of spectroscopic analysis that is used by organic and inorganic chemists is IR spectroscopy. It is simply the measurement of the absorption of different IR frequencies by a sample that is placed in the path of the IR beam in order to measure the absorption. In the IR spectroscopic analysis, one of the key objectives is to determine which chemical functional groups are present in the sample. IR radiation is absorbed at characteristic frequencies by different functional groups. The IR spectrometer is capable of accepting a wide range of samples, including gases, liquids, and solids, using various sampling accessories that can be attached to the device. Consequently, IR spectroscopy has emerged as a very important and popular tool for the discovery of structural characteristics of compounds and identifying them (Chabal, 1988). Figure 11.5 illustrates a schematic diagram of IR spectrometer instrument.

Among the electromagnetic spectrum, IR radiation covers a segment whose wavelength ranges from 0.78 µm up to 1,000 µm at various wavelengths. At high frequencies, it is bound by the red end of the visible region, and at low frequencies, it is bound by the microwave region. This region of the IR spectrum is usually considered to have three categories: near infrared (0.78 to 2.5 µm), mid infrared (2.5 to 50 µm), and far infrared (50 to 1000 µm). The mid-IR region is the most frequently used range (B. Stuart, 2000).

In order to achieve a far IR spectrum, special optical materials and sources must be used. This method is generally used for the analysis of organic, inorganic, and organometallic compounds that contain heavy atoms (mass numbers of more than 19). It is capable of providing valuable information to structural studies of samples such as conformation and lattice dynamics. In order to perform near-IR spectroscopy, very little or no sample preparation is required. The system provides high-speed quantitative analysis without consuming or damaging the sample in the process. For remote analysis, its instruments are often paired with UV-visible spectrometers and

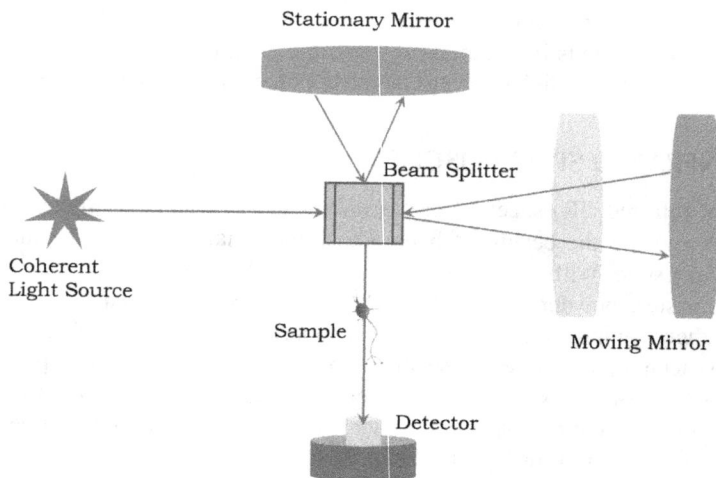

FIGURE 11.5   Schematic diagram of IR spectrometer instrument.

coupled with fiber-optic devices for remote measurement. The popularity of the use of near-IR spectroscopy in process control applications has grown in recent years.

Basically, the IR spectra are obtained by observing changes in transmittance (or absorption) intensity with frequency as a function of changes in transmittance (or absorption). The majority of commercial instruments use dispersive spectrometers or Fourier transform (FT) spectrometers in order to separate and measure IR radiation (B. H. Stuart, 2004).

11.5.1 DISPERSIVE SPECTROMETERS

It should be noted that dispersive spectrometers, which were introduced in the mid-1940s and have become widely used ever since, provided a robust instrumentation that allowed extensive application of the technique to be done. There are three basic components that make up an IR spectrometer: the radiation source, the monochromator, and the detector.

It is common to use an inert solid that is heated electrically to 1,000–1,800 °C as a source of radiation for an IR spectrometer. Monochromators are devices that disperse a wide spectrum of electromagnetic energy and provide a continuous calibrated series of electromagnetic energy bands of determinable wavelengths or frequency ranges to provide continuous electromagnetic energy dispersion. The most common types of detectors used in dispersive IR spectrometers can be categorized into two classes: thermal detectors and photon detectors. There are various types of thermal detectors, such as thermocouples, thermistors, and pneumatic devices (Golay detectors).

11.5.2 SPECTROSCOPIC DESIGN

There are several types of dispersive IR spectrometers that are used today. The typical dispersive type of spectrometer consists of a broad-band source that passes through the sample and is dispersed by a monochromator into component frequencies. When the beams collide with the detector, an electrical signal will be generated, which is recorded by the recorder.

11.5.3 FOURIER TRANSFORM SPECTROMETERS

The use of IR spectroscopy can be found in many different scientific fields. In fact, some of the applications of this technique can be found in organic and inorganic molecules, polymers, biological applications, as well as industrial applications. Due to their superior speed and sensitivity, FT spectrometers have recently replaced dispersive instruments for most of the applications in which they are useful. These advances have greatly enhanced the capabilities of IR spectroscopy and have enabled its use to be applied to a wide range of areas that would otherwise be virtually impossible to analyze with dispersive instruments.

11.5.4 SPECTROMETER DESIGN

It is important to remember that in an FT system, there are three basic components: the radiation source, the interferometer, and the detector. Both dispersive spectrometers and FT spectrometers use the same types of radiation sources to achieve their measurement results. However, it is more common for the Fourier-transform infrared spectroscopy (FTIR) sources to be cooled by water rather than air in order to deliver better power and stability. This process is performed by replacing the monochromator with an interferometer, which divides radiant beams into separate beams, generates an optical path difference between these beams, and then recombines them to produce repeated interference signals, which are measured as a function of the optical path difference by a detector. This type of interferometer produces interference signals, which are a result of the IR spectral information that is generated after a sample passes through the device. A Michelson interferometer is one of the most common types of interferometer used today. This apparatus is composed of three active components: a moving mirror, a fixed mirror, and a beam splitter in order to receive the light (Ferrari et al., 2004).

11.5.5 ADVANTAGES OF FTIR

In comparison with dispersive spectrometers, FTIR has a number of distinct advantages:

Faster and more sensitive. During a single scan of the moving mirror, a complete spectrum of frequencies can be observed, while all frequencies are being observed by the detector simultaneously at the same time.

Optimum throughput. It should be noted that since the interferometer does not require dispersion or filtering, there is no need for energy-consuming slits since there is no need for energy-wasting slits to be used. As an alternative to rectangular apertures in FTIR systems, circular apertures are also commonly used.

Laser internal reference. Most FTIR systems utilize a helium neon laser to provide an internal reference for calibration in order to achieve an accuracy of more than 0.01 to 1 cm by making use of the laser as an internal reference. As a result, there is no need for external calibrations to be carried out.

Design is simpler. A moving mirror is the only part that moves, so there is less wear and better reliability than a mechanical mechanism with multiple moving parts.

Eliminating stray light and emission contributions. There are many frequencies that can be modulated by the interferometer in FTIR. Neither unmodulated stray light nor any unspecified sample emissions are detected by this instrument.

High-powered data station. There is usually a powerful, computerized data system included with modern FTIR spectrometers that produces excellent results. There are various data processing tasks it can perform, such as Fourier transformations, interactive spectral subtraction, baseline correction, smoothing, integration, and searching through a collection of library data.

11.6 CIRCULAR DICHROISM

There is a spectroscopic method known as circular dichroism (CD) that relies on the fact that certain molecules react differently with circularly polarized light that is directed either left or right. It is known that circularly polarized light occurs in two nonsuperimposable forms, namely its mirror images, each of which is the mirror image of the other.

There must be a chiral molecule, such as those found in the vast majority of biological molecules, to be able to distinguish between the two chiral forms of light. In order for a method to be able to recognize subtle differences between molecules (enantiomers) that do not superimpose mirror images (enantiomers), it needs to be highly sensitive to all the three-dimensional properties of the molecules, that is, to their conformation. The CD spectrum of proteins and/or nucleic acids can also be altered by the binding of ligands or by interactions between proteins and/or proteins and DNA. In order to determine chirality in molecules, CD is an essential analytical technique that analyzes the optical activity of molecules to determine their chirality (Rodger & Nordén, 1997).

It is possible to determine equilibrium constants of conformation by examining these changes in CD, as well as providing evidence for conformational changes by examining these changes in CD. Hence, CD can provide information regarding the secondary structural information of proteins and nucleic acids, as well as information regarding the binding of ligands to these types of macromolecules in the secondary structure (Woody, 1995). It is a widely used technique for the elucidation of macromolecular structure, especially proteins and nucleic acids, but it has gained prominence in the scientific community in recent years for its application to a wide variety of molecular structures. An optically active molecule in the vicinity of the chromophore may be preferentially absorbed by circularly polarized light in its vicinity. In the case of optically active (chiral) materials, CD signals are observed. However, in some cases, chirality can also be introduced, for instance, through covalent bonding to a chiral chromophore, or when the chromophore is exposed to an asymmetric environment. A CD technique is commonly used in order to analyze proteins for conformation, to analyze biomolecules for structure, and to identify chiral patterns.

11.6.1 INSTRUMENTATION

Photoelastic modulator (PEM): The purpose of a CD spectrophotometer is to create linearly polarized light in order to produce circularly polarized light. A polarizer is an optical device that aligns the crystal axes and molecule orientations in order to make the unpolarized light pass through it. The linearly polarized light is then converted into circularly polarized light using a device known as a PEM. CD is a type of absorption technique which is based on Beer's law.

$$A = elc$$

where A is the absorbance, e is the molar absorptivity constant, l is the length of the cell path, and c is the concentration of the chromophore. In this regard, it is vital to obtain accurate CD data in order to determine the amount of light that was absorbed by the sample based on concentration and path length. An optimal S/N is obtained when the optical density of the wavelength at which the maximum absorption occurs is equal to ~1. In most cases, protein CD method is used to determine whether a purified, expressed protein is folded, whether a mutation affects its conformation or stability, or whether it is affected by a mutation. Additionally, it can also be used to study the interactions between proteins. There are some basic steps involved in obtaining and interpreting CD data, as well as methods for analyzing spectra to estimate the secondary structural composition of proteins, which are explained in the protocol by Greenfield (Greenfield, 2006).

By utilizing a synchrotron radiation circular dichroism (SRCD) spectroscopy as opposed to a xenon arc lamp as the light source, SRCD spectroscopy extends the range of conventional CD spectroscopy (using laboratory-derived instruments). SRCD spectroscopy uses the intense beam produced by a synchrotron instead of a xenon arc lamp. An SRCD beamline produces a greater number of light pulses (up to four orders of magnitude) than an equivalent conventional CD instrument, plus it covers a wide wavelength range, which includes the vacuum and far- and near-ultraviolet spectrums (e.g., 130–300 nm) for example (Wallace, 2013).

SRCD spectroscopy allows the collection of higher quality and larger amounts of data; therefore, it can be used to enhance and improve the analytical process performed based on the data obtained. By using this method, there are a number of advantages that could be gained, such as the ability to collect data which is normally not achievable using conventional CD instruments. Therefore, a higher level of information content can be obtained in the spectra as a result of being able to measure additional electronic transitions. As a result, it is possible to detect spectra with a higher signal-to-noise ratio than previously, which could allow us to use fewer data sets or shorter collection periods for data collection, or perhaps it will be able to detect subtle differences in the two samples that weren't noticed before. In the research field of protein conformational studies and their complexes, this newly developed method is playing a major role, both in terms of static and dynamic studies. In spite of the limitations of conventional CD spectroscopy, various types of new kinds of experiments have been possible using SRCD spectroscopy, which were not feasible with conventional CD spectroscopy (Miles et al., 2007).

11.7 X-RAY MICROPROBE

There are many advantages of using X-rays for micro probing and characterization of materials since they have a unique interaction with matter compared to electrons and other charged particles. In contrast to regular light, X-rays are more efficient at photo ejecting electrons from the inner shell of the nucleus, resulting in characteristic X-ray fluorescence. Furthermore, X-rays are less likely to cause Bremsstrahlung [Braking radiation] in a fluorescent signal-to-background ratio that is higher than that of electrons because they have a lower bremsstrahlung effect (Sparks & Ice, 1990).

An application of X-ray spectroscopy whereby the sample is bombarded with a finely focused beam of X-rays (approximately 10 m in diameter) to precisely measure the sample's X-ray absorption coefficient. When compared with other methods such as electron probes, proton probes, and ion probes, X-ray elemental analysis using a synchrotron source has many advantages (Winick & Doniach, 2012), such as improved sensitivity and signal-to-background ratio, ease of operation, and the ability to be used in a variety of environments. In terms of the capability of a microprobe, there are three fundamental aspects to be taken into consideration: (a) the energy distribution of the excitation radiation, (b) the intensity of the flux measured by photons, and (c) the spatial resolution of the probe.

In synchrotron radiation sources that use bending magnets or wigglers, the energy distribution of X-rays is a continuous spectrum of X-rays that, depending on the parameters of the synchrotron radiation source, can have a useful X-ray flux up to over 100 keV. It is possible to produce various types of energy distributions from this source by using mirrors, filters, and crystal monochromators (Penner-hahn & Peariso, 2000).

As X-ray microprobes have been designed with large mirrors or crystals as focal points, many geometries have been proposed for the focusing of beams. As a result of the difficulty in fabricating these components, only a small number have been built. Furthermore, the majority of the designs do not contain sufficient demagnification to leave a beam spot size no larger than 50 µm in diameter without the use of a pinhole as a demagnification device (Bertsch & Hunter, 2001; Winick & Doniach, 2012). There is great potential for the X-ray microprobe to be used in a wide range of applications, including biological studies (Wu et al., 1990).

In order to analyze trace elements at micrometer spatial resolutions using synchrotron radiation (SR), an X-ray microprobe is currently being developed. Under the beamlines, a focusing mirror and an ellipsoidal mirror are used for this purpose. Presently, white light is being used as an excitation source for the characteristic X-ray fluorescence lines to be observed (Bertsch & Hunter, 2001). As a result of advanced synchrotron-based microprobe techniques and data analysis methods, researchers are now able to investigate the spatial variation and source of heavy metals at multiple scales at once (Lombi & Susini, 2009). Yu et al. (2020) used synchrotron-based microprobe techniques and advanced data analytical methods to probe into elemental distribution and quantify the iron species at microscales.

11.8 SUMMARY

In order to provide a better understanding of the subject matter, this chapter has provided a general overview of synchrotron techniques and their application in spectroscopy. A comprehensive overview of many aspects of spectroscopy was provided, from the origins of the field to the development of a wide range of new fields. Over the past few decades, there has been a variety of advancements in the field of material science and measurement as a result of various advancements occurred in a wide range of areas. This chapter focused on this topic by discussing the various types of synchrotron technologies, their applications in materials research, analysis, using spectroscopy techniques, as well as the advancement in their application of these technologies in the modern world.

REFERENCES

Bertsch, P. M., & Hunter, D. B. (2001). Applications of synchrotron-based X-ray microprobes. *Chemical Reviews*, *101*(6), 1809–1842. https://doi.org/10.1021/cr990070s

Chabal, Y. J. (1988). Surface infrared spectroscopy. *Surface Science Reports*, *8*(5–7), 211–357.

Clegg, W. (2000). Synchrotron chemical crystallography. *Journal of the Chemical Society, Dalton Transactions, 19*, 3223–3232. https://doi.org/10.1039/B004136J

Czajka, A., & Armes, S. P. (2021). Time-resolved small-angle X-ray scattering studies during aqueous emulsion polymerization. *Journal of the American Chemical Society, 143*(3), 1474–1484. https://doi.org/10.1021/jacs.0c11183

de Groot, F. (2001). High-resolution X-ray emission and X-ray absorption spectroscopy. *Chemical Reviews, 101*(6), 1779–1808. https://doi.org/10.1021/cr9900681

Ferrari, M., Mottola, L., & Quaresima, V. (2004). Principles, techniques, and limitations of near infrared spectroscopy. *Canadian Journal of Applied Physiology, 29*(4), 463–487.

Grabowski, M., Cooper, D. R., Brzezinski, D., Macnar, J. M., Shabalin, I. G., Cymborowski, M., Otwinowski, Z., & Minor, W. (2021). Synchrotron radiation as a tool for macromolecular X-ray crystallography: A XXI century perspective. *Nuclear Instruments and Methods in Physics Research Section B: Beam Interactions with Materials and Atoms, 489*, 30–40. https://doi.org/https://doi.org/10.1016/j.nimb.2020.12.016

Green, J. C. (2007). 1.14–Photoelectron spectroscopy. In D. M. P. Mingos & R. H. Crabtree (Eds.), *Comprehensive Organometallic Chemistry III* (pp. 381–406). Elsevier. https://doi.org/https://doi.org/10.1016/B0-08-045047-4/00015-7

Greenfield, N. J. (2006). Using circular dichroism spectra to estimate protein secondary structure. *Nature Protocols, 1*(6), 2876–2890. https://doi.org/10.1038/nprot.2006.202

Hogg, T., & Hilgenfeld, R. (2007). 3.38–Protein crystallography in drug discovery. In J. B. Taylor & D. J. Triggle (Eds.), *Comprehensive Medicinal Chemistry II* (pp. 875–900). Elsevier. https://doi.org/https://doi.org/10.1016/B0-08-045044-X/00111-5

Holmes, J. (2017). Photoelectron spectroscopy. In J. C. Lindon, G. E. Tranter, & D. W. Koppenaal (Eds.), *Encyclopedia of Spectroscopy and Spectrometry (Third Edition)* (p. 618). Academic Press. https://doi.org/https://doi.org/10.1016/B978-0-12-803224-4.00355-1

Hüfner, S. (2003). Introduction and basic principles. In *Photoelectron Spectroscopy* (pp. 1–60). Springer. https://doi.org/10.1007/978-3-662-09280-4 eBook ISBN 978-3-662-09280-4

Joosten, R. P., Chinea, G., Kleywegt, G. J., & Vriend, G. (2007). 3.23–Protein three-dimensional structure validation. In J. B. Taylor & D. J. Triggle (Eds.), *Comprehensive Medicinal Chemistry II* (pp. 507–530). Elsevier. https://doi.org/https://doi.org/10.1016/B0-08-045044-X/00096-1

Law, J. (1973). The development of specialties in science: The case of X-ray protein crystallography. *Science Studies, 3*(3), 275–303. https://doi.org/10.1177/030631277300300303

Li, T., Senesi, A. J., & Lee, B. (2016). Small angle X-ray scattering for nanoparticle research. *Chemical Reviews, 116*(18), 11128–11180. https://doi.org/10.1021/acs.chemrev.5b00690

Liu, X. C., Ge, X. G., Li, Y. F., An, X. M., Jiang, L., Guo, H., Sun, Z. L., Miao, X. R., & Lu, F. X. (2022). Preparation of single-crystal diamond for small angle X-ray scattering in situ loading test. *Diamond and Related Materials, 121*, 108719. https://doi.org/https://doi.org/10.1016/j.diamond.2021.108719

Lokteva, I., Dartsch, M., Dallari, F., Westermeier, F., Walther, M., Grübel, G., & Lehmkühler, F. (2021). Real-time X-ray scattering discovers rich phase behavior in PbS nanocrystal superlattices during in situ assembly. *Chemistry of Materials, 33*(16), 6553–6563. https://doi.org/10.1021/acs.chemmater.1c02159

Lombi, E., & Susini, J. (2009). Synchrotron-based techniques for plant and soil science: Opportunities, challenges and future perspectives. *Plant and Soil, 320*(1), 1–35. https://doi.org/10.1007/s11104-008-9876-x

Miles, A. J., Hoffmann, S. V., Tao, Y., Janes, R. W., & Wallace, B. A. (2007). Synchrotron radiation circular dichroism (SRCD) spectroscopy: New beamlines and new applications in biology. *Spectroscopy*, *21*(5–6), 245–255.

Moreno, A. (2017). Advanced methods of protein crystallization. In A. Wlodawer, Z. Dauter, & M. Jaskolski (Eds.), *Protein Crystallography: Methods and Protocols* (pp. 51–76). Springer. https://doi.org/10.1007/978-1-4939-7000-1_3

Myers, N. (2008). Molecular embodiments and the body-work of modeling in protein crystallography. *Social Studies of Science*, *38*(2), 163–199. https://doi.org/10.1177/0306312707082969

Natoli, C. R., Benfatto, M., della Longa, S., & Hatada, K. (2003). X-ray absorption spectroscopy: State-of-the-art analysis. *Journal of Synchrotron Radiation*, *10*(1), 26–42.

Ng, L. M., & Simmons, R. (1999). Infrared spectroscopy. *Analytical Chemistry*, *71*(12), 343–350.

Penner-Hahn, J. E. (2003). X-ray absorption spectroscopy. *Comprehensive Coordination Chemistry II*, *2*, 159–186.

Penner-hahn, J. E., & Peariso, K. (2000). X-ray microprobe imaging and X-ray microspectroscopy in biology. *Synchrotron Radiation News*, *13*(5), 22–30. https://doi.org/10.1080/08940880008261097

Raschpichler, C., Goroncy, C., Langer, B., Antonsson, E., Wassermann, B., Graf, C., Klack, P., Lischke, T., & Rühl, E. (2020). Surface properties and porosity of silica particles studied by wide-angle soft X-ray scattering. *The Journal of Physical Chemistry C*, *124*(30), 16663–16674. https://doi.org/10.1021/acs.jpcc.0c04308

Reinert, F., & Hüfner, S. (2005a). Photoemission spectroscopy—from early days to recent applications. *New Journal of Physics*, *7*, 97–97. https://doi.org/10.1088/1367-2630/7/1/097

Reinert, F., & Hüfner, S. (2005b). Photoemission spectroscopy—from early days to recent applications. *New Journal of Physics*, *7*(1), 97.

Rodger, A., & Nordén, B. (1997). *Circular Dichroism and Linear Dichroism* (Vol. 1). Oxford University Press.

Rotella, F. J., Alkire, R. W., Duke, N. E. C., & Molitsky, M. J. (2011). Diagnostic tools used in the calibration and verification of protein crystallography synchrotron beam lines and apparatus. *Nuclear Instruments and Methods in Physics Research Section A: Accelerators, Spectrometers, Detectors and Associated Equipment*, *649*(1), 228–230. https://doi.org/https://doi.org/10.1016/j.nima.2010.12.193

Sauer, K., Yano, J., & Yachandra, V. K. (2005). X-ray spectroscopy of the Mn4Ca cluster in the water-oxidation complex of photosystem II. *Photosynthesis Research*, *85*(1), 73–86.

Sauer, K., Yano, J., & Yachandra, V. K. (2008). X-ray spectroscopy of the photosynthetic oxygen-evolving complex. *Coordination Chemistry Reviews*, *252*(3), 318–335. https://doi.org/https://doi.org/10.1016/j.ccr.2007.08.009

Sparks, C. J., & Ice, G. E. (1990). X-ray microprobe-microscopy. *AIP Conference Proceedings*, *215*(1), 770–786. https://doi.org/10.1063/1.39851

Stuart, B. (2000). Infrared spectroscopy. *Kirk-Othmer Encyclopedia of Chemical Technology*. John Wiley & Sons, Inc. 1–20. https://doi.org/10.1002/0471238961.0914061810151405.a01.pub3

Stuart, B. H. (2004). *Infrared Spectroscopy: Fundamentals and Applications*. John Wiley & Sons.

Stubbs, M. T. (2007). Protein crystallography. In J. B. Taylor & D. J. Triggle (Eds.), *Comprehensive Medicinal Chemistry II* (pp. 449–472). Elsevier. https://doi.org/https://doi.org/10.1016/B0-08-045044-X/00094-8

Wallace, B. A. (2013). Synchrotron radiation circular dichroism spectroscopy. In G. C. K. Roberts (Ed.), *Encyclopedia of Biophysics* (pp. 2559–2560). Springer. https://doi.org/ 10.1007/978-3-642-16712-6_644

Wende, H. (2004). Recent advances in x-ray absorption spectroscopy. *Reports on Progress in Physics, 67*(12), 2105.

Winick, H., & Doniach, S. (2012). *Synchrotron Radiation Research.* Springer Science & Business Media.

Woody, R. W. (1995). [4] Circular dichroism. *Methods in Enzymology, 246*, 34–71. https://doi. org/https://doi.org/10.1016/0076-6879(95)46006-3

Wu, Y., Thompson, A. C., Underwood, J. H., Giauque, R. D., Chapman, K., Rivers, M. L., & Jones, K. W. (1990). A tunable X-ray microprobe using synchrotron radiation. *Nuclear Instruments and Methods in Physics Research Section A: Accelerators, Spectrometers, Detectors and Associated Equipment, 291*(1), 146–151. https://doi.org/https://doi.org/ 10.1016/0168-9002(90)90050-G

Xiao, Y., & Lu, X. (2019). Morphology of organic photovoltaic non-fullerene acceptors investigated by grazing incidence X-ray scattering techniques. *Materials Today Nano, 5*, 100030. https://doi.org/https://doi.org/10.1016/j.mtnano.2019.100030

Yano, J., & Yachandra, V. K. (2009). X-ray absorption spectroscopy. *Photosynthesis Research, 102*(2), 241–254.

Yu, X., Wang, Y., & Lu, S. (2020). Tracking the magnetic carriers of heavy metals in contaminated soils based on X-ray microprobe techniques and wavelet transformation. *Journal of Hazardous Materials, 382*, 121114.

12 Other Advanced Instruments Used for Characterization of Functionally Graded Materials

12.1 NEAR-EDGE X-RAY ABSORPTION FINE STRUCTURE (NEXAFS)

(NEXAFS is a spectroscopy technique designed to study the fine structure of the absorption close to the absorption edge, which is around 30eV above the actual absorption edge. An X-ray absorption near edge structure (XANES), also referred to as NEXAFS, is a method of analyzing spectra produced by X-ray absorption spectroscopy in an effort to obtain a more detailed description. This type of spectroscopic analysis determines the partial density of the empty states of a molecule by using an element-specific and local bonding-sensitive spectroscopic method.

NEXAFS is also very sensitive to the bonding environment that surrounds the atom that is absorbing light. Over the elemental absorption edges in the NEXAFS spectrum, there is a considerable amount of fine structure. As a result of excitations into unoccupied molecular orbitals, this fine structure is produced. In comparison to extended X-ray absorption fine structure (EXAFS), which is also an extremely weak structure in polymers, this structure has a much larger area than the higher energy EXAFS.

It has been determined that EXAFS is primarily caused by scattered photoelectron events generated by high-energy photoelectron photon collisions of neighboring atomic cores, while NEXAFS is primarily caused by multiple scattering of a low-energy photoelectron in a valence potential created by several neighbors. In many cases, it is possible to identify the local bonding environment through the use of spectral "fingerprinting" techniques.

NEXAFS spectroscopy has another great advantage over other spectroscopy techniques in that it is dependent on polarization. For covalent systems such as molecules with a low z-factor, macromolecules or polymers, which possess directional bonds, linearly polarized X-rays are the best solution. This can be viewed as an example of an X-ray being able to locate chemical bonds in the atom by the direction of its electric field vector, which can be viewed as a "search light" that can investigate the direction of a chemical bond located at the absorption edge of an atom.

DOI: 10.1201/9781003340546-13

NEXAFS is a type of absorption spectroscopy. In order for NEXAFS to function, an X-ray photon must be absorbed by a core level of an atom in a solid followed by its emission as a photoelectron as a result of that absorption. As a result of this process, the resulting core hole is filled either by means of an Auger process or by the capture of an electron from another shell, followed by the emission of a fluorescent photon. In terms of the bonding environment of the atom that absorbs light, NEXAFS is extremely sensitive to that environment. In many cases, it is possible to use spectral "fingerprint" techniques in order to locate the location of the bonding environment (Nascimento et al., 2020).

It should be noted that a core hole is the space occupied by a core electron before it absorbs an X-ray photon and ejected from its shell, as stated above. There are a number of factors that contribute to the instability of core holes, such as their high energy. The average life span of a core hole is close to one femtosecond, which is a fairly short amount of time. It is believed that they are formed by processes in which either a core electron absorbs an X-ray photon (X-ray absorption) or absorbs part of the kinetic energy of an X-ray photon (X-ray Raman scattering). A successor process to the Auger electron ejection is the decay of the core hole which may take place either through X-ray fluorescence or through the Auger electron ejection.

In the case of carbon-containing substances, NEXAFS spectroscopy is an effective technique for determining the electronic structure, the symmetry, and the energy distribution of the unoccupied electron states as well as how the atoms are arranged in chemistry and in the bonding between them (Stöhr, 1992). In NEXAFS spectroscopy, soft X-rays are used to probe the electronic transitions from core levels to higher lying unoccupied states at the same time. It is important to note that the energy levels of both the initial and final states are affected by the molecular bonds and their chemical environment, which is why the spectral features of the near-edge fine structure provide a unique "molecular fingerprint." As a result, NEXAFS spectroscopy has become a well-established analytical method for the analysis of compositional surfaces and is also one of the most important applications of synchrotron radiation today (Kühl et al., 2016).

There is one major obstacle that is common to all NEXAFS experiments and this problem is the fact that the mean free path of soft X-rays in air is only a few millimeters, which necessitates the use of a high vacuum environment for the source, sample, and spectrometer. In addition to outgassing problems when aqueous solutions or adsorbed gases are to be analyzed, water dehydration can also result in structural changes in the sample, such as proteins, deoxyribonucleic acid, and transition metal compounds that are hydrated. As a consequence of drying, covalent bonds can be formed between substrate and adsorbate which can result in a change in the signal produced by NEXAFS. Synchrotrons employ specially designed gas cells for in situ measurements as well as helium-purged sample compartments for reducing the incidental loss of helium, since helium at atmospheric pressure is by far less absorbing than air at ambient temperatures (Drisdell & Kortright, 2014; Smith et al., 1995).

This means that NEXAFS can provide a sensitive probe to detect subtle geometric differences that are induced by differences in the chemical reactions that have resulted in the formation of the polymers and the differences in their properties. It has been reported that NEXAFS provides one of the most accurate experimental measurements of molecular orientation despite its difficulties with background subtraction and step normalization (Duncan, 2018).

In their work, Xu et al. (2019) demonstrated direct evidence of Fe^{3+} and Fe^{2+} gradient distribution across cuticle thickness, which shows a higher concentration of Fe^{2+} inside the inner cuticle, which supports the hypothesis that the cuticle is a functionally graded material with high stiffness, extensibility, and self-healing abilities. The authors examined the oxidation states of iron ions in the cuticle, a fresh thread was collected and analyzed using XANES and X-ray photoelectron spectroscopy based on C K-edge absorption near-edge spectroscopy. A significant contribution to the understanding of the mechanisms that result in the strength and self-healing of mussel cuticles is expected to be gained by the findings of this study, which could have implications for the design and fabrication of bioinspired materials that are tough, strong, and self-healing.

12.2   EXTENDED X-RAY ABSORPTION FINE STRUCTURE

EXAFS and XANES are two regions of the spectrum obtained from X-ray absorption spectroscopy. There has been a large number of EXAFS studies that have indicated an oscillating part of the spectrum to the right of the absorption edge (appearing as a sudden sharp peak), starting at approximately 50 eV and extending just above 1,000 eV above the absorption edge. In recent years, EXAFS has become a widely used tool for providing information about the local environment of an atom, with applications extending well beyond the geometric analysis of amorphous crystalline solids. EXAFS has become increasingly popular to analyze nanocrystalline materials quantitatively owing to its significant application. It is important to realize that when studying a single atom within a material, a number of properties are analyzed, including its coordination number, its disorder, and the distance between its neighboring atoms. The more precise the structural information obtained theoretically is, the further down the EXAFS region you go, approximately 0.02 Angstroms or more (Lee et al., 1981).

It has been demonstrated that XAFS experiments with EXAFS as well as XANES experiments can be used with great success to characterize highly dispersed supported metals with fine structures. Using EXAFS with a supported structure such as an alumina surface to measure the local structure of highly dispersed metal on a surface, such as alumina, in conjunction with XANES to measure the valence state of the metal on the surface, will also provide useful information about its local structure. It has been observed that oscillations can be spotted in the photon energy spectrum just above an element's X-ray absorption edge with EXAFS (Stern, 1974).

It has been also demonstrated through theoretical studies of these oscillations that they contain a number of structural clues about the immediate environment of the atom undergoing X-ray absorption in which they are observed. As a result of the absorption of an X-ray photon by an atom, the atom emits a photoelectron. It is possible that the photoelectron wave emitted in the outgoing direction may be reflected by the neighboring atoms. The oscillations observed just above the edge of X-ray absorption are caused by the constructive and destructive interference between the outgoing and reflected photoelectron waves. As a result of the Fourier transform of the oscillations, structural information is extracted. It can be observed that absorption spectra are influenced by nearest-neighbor distances in terms of peak intensity. An oscillation spectrum of a bulk material with a known nearest-neighbor distance is used to calibrate the abscissa based on the oscillation spectrum of a reference material. A high level of precision can be obtained in the measurement of distances to neighboring atoms, and because of this, these distances can be associated with the type of bonding, such as metal–metal or metal–oxygen. Based on the intensity of the absorption peaks, one can deduce the number of nearest and next nearest neighboring atoms based on their distance from each other (Farrauto & Hobson, 2003).

Ekwongsa et al. (2020) reported temperature-dependent local structure of $LiCoO_2$ determined by in situ Co K-edge EXAFS. A change in local structure depending on temperature around Co atoms was reported (Ekwongsa et al., 2020). Using chemical vapor deposition techniques, Parida et al. (2021) investigated the local structure of GaN nanowires (NWs) synthesized by the chemical vapor deposition method with different native defect concentrations. EXAFS was used to address the effects of dopant incorporation as well as other defects on the values of the coordination number and bond length in this study. According to the local tetrahedral structure of GaN, the decrease in bond length values along preferential crystal axes emphasizes the preferred site for oxygen doping within the lattice of the material. The EXAFS analysis was used to investigate the changes in the local tetrahedral structure of GaN with the incorporation of Al. This analysis provides a clear understanding of how to choose a suitable process for the ternary III-nitride random alloy formation process to be applied (Parida et al., 2021).

The study by Ju et al. (2020) examined the electrode stabilization due to surface passivation in order to develop long-cycle lithium-ion batteries (LIBs). It has been demonstrated in this work that functionally graded materials consisting of a nickel oxyfluoride (NiFeOF) cathode, covered by a homologous passivation layer, are rationally designed for long-cycle LIBs using iron-doped NiFeOF as the cathode material. Due to the fact that X-ray absorption spectroscopy is a surface-sensitive method, this suggests that the surface layer contains Ni that is being oxidized, therefore suggesting that Ni might be bonded to either F or even O, as confirmed by EXAFS. The EXAFS fitting according to a model indicates the existence of NiF_x as a component of the system. XANES and EXAFS both show a strong contribution from the Ni metal, which is reflected by both the analyses (Ju et al., 2020).

12.3 SUMMARY

The purpose of this chapter is to provide an overview of the different types of absorption spectroscopy that can be used to diagnose functional graded materials in their X-ray absorption spectrums for transitions from core states to higher energy states. In a number of scientific disciplines, the utilization of XAFS spectroscopy at the Synchrotron is now widely accepted as a mainstream technique that can provide molecular level information. The XAFS spectrum has two distinct regions of fine structure, known as the near edge and the extended edge. The two techniques were discussed and an overview of both of them was presented.

REFERENCES

Drisdell, W. S., & Kortright, J. B. (2014). Gas cell for in situ soft X-ray transmission-absorption spectroscopy of materials. *Review of Scientific Instruments*, *85*(7), 074103.

Duncan, D. A. (2018). Synchrotron-based spectroscopy in on-surface polymerization of covalent networks. In K. Wandelt (Ed.), *Encyclopedia of Interfacial Chemistry* (pp. 436–445). Elsevier. https://doi.org/https://doi.org/10.1016/B978-0-12-409547-2. 13768-0

Ekwongsa, C., Rujirawat, S., Butnoi, P., Vittayakorn, N., Suttapun, M., Yimnirun, R., & Kidkhunthod, P. (2020). Temperature dependent local structure of $LiCoO_2$ determined by in-situ Co K-edge X-ray absorption fine structure (EXAFS). *Radiation Physics and Chemistry*, *175*, 108545. https://doi.org/https://doi.org/10.1016/j.radphyschem. 2019.108545

Farrauto, R. J., & Hobson, M. C. (2003). Catalyst characterization. In R. A. Meyers (Ed.), *Encyclopedia of Physical Science and Technology (Third Edition)* (pp. 501–526). Academic Press. https://doi.org/https://doi.org/10.1016/B0-12-227410-5/00087-9

Ju, L., Wang, G., Liang, K., Wang, M., Sterbinsky, G. E., Feng, Z., & Yang, Y. (2020). Significantly improved cyclability of conversion-type transition metal oxyfluoride cathodes by homologous passivation layer reconstruction. *Advanced Energy Materials*, *10*(9), 1903333. https://doi.org/https://doi.org/10.1002/aenm.201903333

Kühl, F.-C., Müller, M., Schellhorn, M., Mann, K., Wieneke, S., & Eusterhues, K. (2016). Near-edge x-ray absorption fine structure spectroscopy at atmospheric pressure with a table-top laser-induced soft x-ray source. *Journal of Vacuum Science & Technology A*, *34*(4), 041302. https://doi.org/10.1116/1.4950599

Lee, P. A., Citrin, P. H., Eisenberger, P. t., & Kincaid, B. M. (1981). Extended x-ray absorption fine structure—its strengths and limitations as a structural tool. *Reviews of Modern Physics*, *53*(4), 769.

Nascimento, D. R., Zhang, Y., Bergmann, U., & Govind, N. (2020). Near-edge X-ray absorption fine structure spectroscopy of heteroatomic core-hole states as a probe for nearly indistinguishable chemical environments. *The Journal of Physical Chemistry Letters*, *11*(2), 556–561.

Parida, S., Sahoo, M., Tromer, R. M., Galvao, D. S., & Dhara, S. (2021). Effect of oxygen and aluminum incorporation on the local structure of GaN nanowires: Insight from extended X-ray absorption fine structure analysis. *The Journal of Physical Chemistry C*, *125*(5), 3225–3234. https://doi.org/10.1021/acs.jpcc.0c10669

Smith, A. D., Derbyshire, G. E., Farrow, R. C., Sery, A., Raudorf, T. W., & Martini, M. (1995). A solid state detector for soft energy extended x-ray absorption fine structure measurements. *Review of Scientific Instruments*, *66*(2), 2333–2335. https://doi.org/ 10.1063/1.1145680

Stern, E. A. (1974). Theory of the extended x-ray-absorption fine structure. *Physical Review B*, *10*(8), 3027.

Stöhr, J. (1992). *NEXAFS Spectroscopy* (Vol. 25). Springer Science & Business Media.

Xu, Q., Xu, M., Lin, C.-Y., Zhao, Q., Zhang, R., Dong, X., Zhang, Y., Tian, S., Tian, Y., & Xia, Z. (2019). Metal coordination-mediated functional grading and self-healing in mussel byssus cuticle. *Advanced Science*, *6*(23), 1902043. https://doi.org/https://doi. org/10.1002/advs.201902043

For Product Safety Concerns and Information please contact our EU
representative GPSR@taylorandfrancis.com
Taylor & Francis Verlag GmbH, Kaufingerstraße 24, 80331 München, Germany

www.ingramcontent.com/pod-product-compliance
Lightning Source LLC
Chambersburg PA
CBHW052014230326
41598CB00078B/3421

* 9 7 8 1 0 3 2 3 7 5 1 1 3 *